Repairing Domestic Climate Displacement

Climate change, sometimes thought of as a problem for the future, is already impacting people's lives around the world: families are losing their homes, lands and livelihoods as a result of sea level rise, increased frequency and intensity of storms, drought and other phenomena. Following several years of preparatory work across the globe, legal scholars, judges, UN officials and climate change experts from 11 countries came together to finalise a new normative framework aiming to strengthen the right of climate-displaced persons, households and communities. This resulted in the approval of the Peninsula Principles on Climate Displacement within States in August 2013.

This book provides detailed explanations and interpretations of the Peninsula Principles and includes in-depth discussion of the legal, policy and programmatic efforts needed to uphold the standards and norms embedded in the Principles. The book provides policy-makers with the conceptual understanding necessary to ensure that national-level policies are in place to respond to the climate displacement challenge, as well as a firm sense of the programme-level approaches that can be taken to anticipate, reduce and manage climate displacement. It also provides students and policy advocates with the necessary information to debate and critique responses to climate displacement at different levels.

Drawing together key thinkers in the field, this volume will be of great relevance to scholars, lawyers, legal advisers and policy-makers with an interest in climate change, environmental policy, disaster management and human rights law and policy.

Scott Leckie is the Founder and Director of Displacement Solutions, Visiting Professor at the Australian National University, Canberra and Senior Fellow at Melbourne University Law School, Melbourne, Australia.

Chris Huggins is a member of the Association of American Geographers and is Adjunct Professor at Carleton University, Canada.

Routledge Studies in Development, Displacement and Resettlement

Land Solutions for Climate Displacement
Scott Leckie

Development-Induced Displacement and Resettlement
Edited by Irge Satiroglu and Narae Choi

Resettlement Policy in Large Development Projects
Edited by Ryo Fujikura and Mikiyasu Nakayama

Global Implications of Development, Disasters and Climate Change
Edited by Susanna Price and Jane Singer

Repairing Domestic Climate Displacement
The Peninsula Principles
Edited by Scott Leckie and Chris Huggins

Repairing Domestic Climate Displacement

The Peninsula Principles

Edited by Scott Leckie and Chris Huggins

LONDON AND NEW YORK

First published 2016
by Routledge
2 Park Square, Milton Park, Abingdon, Oxfordshire OX14 4RN

and by Routledge
711 Third Avenue, New York, NY 10017

First issued in paperback 2017

Routledge is an imprint of the Taylor & Francis Group, an informa business

British Library Cataloguing-in-Publication Data
A catalogue record for this book is available from the British Library

Library of Congress Cataloging-in-Publication Data
Repairing domestic climate displacement: the Peninsula Principles /
edited by Scott Leckie and Chris Huggins.
  pages cm
  Includes bibliographical references and index.
  1. Climatic changes–Law and legislation. 2. Environmental refugees–
Legal status, laws, etc. 3. Environmental law, International. I.
Leckie, Scott, editor. II. Huggins, Christopher, editor.
  K3585.5.R468 2016
  344.04′633–dc23   2015012737

ISBN 13: 978-1-138-06498-0 (pbk)
ISBN 13: 978-1-138-92038-5 (hbk)

Typeset in Sabon
by Wearset Ltd, Boldon, Tyne and Wear

For climate-displaced persons everywhere past, present, future

Contents

Contributors

Simon Bagshaw is a Senior Policy Adviser on protection and displacement at the United Nations Office for the Coordination of Humanitarian Affairs. Prior to this, he worked for three years as a senior research associate in the Office of the Representative of the United Nations Secretary-General on Internally Displaced Persons. He holds a PhD in Law from the European University Institute in Florence.

Robin Bronen lives in Alaska, works as a human rights attorney and has been researching 'climigration', the climate-induced relocation of Alaska Native communities since 2007. Her research has been publicised by CNN, the *Guardian* and others. She is a senior research scientist at the University of Alaska Fairbanks and co-founded and works as the executive director of the Alaska Institute for Justice, a non-governmental organisation that is the only immigration legal service provider in Alaska. It houses a Language Interpreter Center, training bilingual Alaskans to be interpreters, and also is a research and policy institute focused on climate justice issues. The Alaska Bar Association awarded her the 2007 Robert Hickerson Public Service award and the 2012 International Human Rights award. The Federal Bureau of Investigation awarded the Alaska Institute for Justice the 2012 FBI Director's Community Service award for its work with human trafficking victims and the International Soroptimists awarded her the 2012 Advancing the Rights of Women award.

Bruce Burson is a senior member of the New Zealand Immigration and Protection Tribunal. He has written a number of the tribunal's leading judgments, including in relation to natural disasters, climate change and international protection law. He is a Senior Research Associate at the Refugee Law Initiative at the University of London, and is a member of the Consultative Committee of the Nansen Initiative on Cross-Border Disaster-Induced Displacement. He is also a member of the International Law Association Committee on the International Law Implications of Sea-level Rise. He is the editor of *Climate Change and Migration: South Pacific Perspectives* (2009) and is co-editor of *On the*

Borders of Refugee Protection? The Impact of Human Rights Law on Refugee Law: Comparative Practice and Theory (Martinus Nijhoff, 2015).

Bonnie Docherty is a lecturer on law and a senior clinical instructor at the International Human Rights Clinic at Harvard Law School. She has done significant fieldwork, scholarship and teaching in the field of human rights and the environment, focusing on climate change and mining. She also has extensive experience with the negotiation and implementation of treaties, particularly in the field of disarmament. She received her J.D. from Harvard Law School and her A.B. from Harvard University.

Khaled Hassine is a lawyer and housing, land and property (HLP) rights specialist. He has worked on HLP issues for over ten years with HLP NGOs, international organisations and research institutes focusing on property and land management, cadastral/titling systems, forced evictions/relocations, security of tenure, and property restitution mechanisms in post-conflict/-disaster settings, including procedural modalities, as well as climate displacement. He has advised numerous UN peace operations in various regions, including Georgia, Sudan, Yemen and Kosovo, and he is a member of the HLP-Group, a Sub-Working Group of the UN Global Protection Cluster, and of the UNHCR Livelihoods Advisory Board. Together with Scott Leckie, he co-authored the first Commentary on the Pinheiro Principles (UN Principles on Housing and Property Restitution for Refugees and Displaced Persons) (forthcoming).

David Hodgkinson is Special Counsel at the national Australian law firm Clayton Utz, Associate Professor in the Law School at the University of Western Australia, and a director of HodgkinsonJohnston Pty Ltd. He practises in the areas of climate change and aviation.

Chris Huggins is a researcher, lecturer and trainer with more than 15 years experience on land and property rights and displacement issues, especially in Sub-Saharan Africa. He is currently a Non-Resident Research Fellow at the African Centre for Technology Studies (ACTS), which was ranked amongst the top 25 most influential climate change think tanks in the world in 2013. He is also an Adjunct Professor at Carleton University, in Ottawa. He has consulted with UN agencies, bilateral donors, policy think tanks, international non-governmental organisations and for-profit institutions. He co-edited (with Scott Leckie) the handbook on *Conflict and Housing, Land and Property Rights* (Cambridge University Press, 2011) which has been included on the syllabi of several university courses, including the University of Edinburgh, University of Tulsa, and University of Copenhagen.

Scott Leckie is an international human rights lawyer and the Director and Founder of Displacement Solutions (www.displacementsolutions.org), an international not-for-profit organisation dedicated to resolving cases of forced displacement throughout the world, in particular displacement caused by climate change and conflict. Throughout his 30-year human ' rights career, he has established several other human rights and political organisations and institutions. He regularly advises several United Nations agencies on various housing, land and property rights issues, and has worked on these questions in 81 countries. He has been active on various international human rights standard-setting initiatives, and was the drafter and driving force behind more than 50 international human rights standards. Scott has written and edited 19 books and over 150 articles and reports on issues including housing rights, economic, social and cultural rights, forced evictions, the right to housing and property restitution for refugees and internally displaced persons and other human rights themes. He lectures frequently and teaches several human rights courses in some of the top 15 law schools around the world, including the world's first law school course on climate change and displacement at the College of Law of the Australian National University and University of Melbourne Law School.

Ezekiel Simperingham is an international lawyer with 12 years of experience providing legal and policy advice on human rights, housing, land and property (HLP), rule of law and displacement (refugee and internally displaced person) issues, and has extensive experience working in conflict- and disaster-affected countries across the globe. Since 2008, he has worked as an international legal consultant for Displacement Solutions, focusing on the design and implementation of legal and policy solutions to situations of displacement globally. He previously worked for the International Commission of Jurists in Bangkok, Thailand, the Office of the High Commissioner for Human Rights in Colombo, Sri Lanka, the International Center for Transitional Justice in New York, USA, the Refugee Status Appeals Authority in Auckland, New Zealand, and the United Nations High Commissioner for Refugees in Canberra, Australia. In 2014, he undertook an emergency deployment as the National Housing, Land and Property Adviser for the Shelter Cluster in the humanitarian response to Typhoon Haiyan in the Philippines. He was awarded a Master of Laws by the New York University School of Law and holds Bachelor of Laws (Hons) and Bachelor of Arts degrees from the University of Auckland, New Zealand. He has authored a number of publications on refugee law, climate change-induced displacement, international criminal justice and HLP.

Foreword

The displacement of people gives rise to acute human rights issues, whether the cause is natural disaster, war or resource development. Various international texts offer guidance as to how such issues can and should be addressed. Until now, however, no set of principles has applied specifically to the displacement of people within states due to climate change. The Peninsula Principles address that subject.

As explained in this introductory book of essays, the normative foundation of the Peninsula Principles is international human rights law, particularly the two international human rights covenants – the International Covenant on Civil and Political Rights and the International Covenant on Economic, Social and Cultural Rights. Drawing on the general content of the human rights relating to home, land and property, the Principles offer considered and comprehensive guidance as to how states and their agencies can effectively meet their obligations to prevent displacement due to climate change and deal in a rights-sensitive manner with its consequences.

In so doing, the Peninsula Principles draw on the now established tradition of principle-making in international law, particularly human rights law. Numerous examples exist of principles being developed through mechanisms that bring together the combined knowledge of international scholars, experts and practitioners. Such principles can provide authoritative guidance in the application and interpretation of human rights standards by states, judiciaries and others. Time should prove the Peninsula Principles to be a welcome addition to this body of soft law.

Particular features of the phenomenon of internal displacement of people due to climate change make the Peninsula Principles especially welcome. History has shown that such displacement has a profoundly negative impact on the lives of individuals, families and (more usually) whole communities. History has also shown that most of this impact is felt by people displaced within states. There is much action that states can – and under international law must – take to prevent or deal rights-sensitively with this problem. States can look to the Peninsula Principles as a framework for that action.

Although a powerful case can be made for the adoption of an international convention dealing specifically with the displacement of people due to climate change (internally or externally), we do not yet have such a convention. Consequently, the two covenants and other general sources of human rights law must carry the immense normative load that is generated by this pressing contemporary problem. It seems very likely that judicial institutions will have to interpret and develop such general standards in a way that meets these particular demands. Like analogous principles in other fields, the Peninsula Principles should prove to be a valuable source in that process.

Justice Kevin Bell
Supreme Court of Victoria
Melbourne, Australia

Preface

Displacement Solutions (DS)[1] believes that the time for concerted action to prevent and resolve climate displacement is upon us, and has been for some time. In Bangladesh, the Solomon Islands, Kiribati, Panama, the US state of Alaska, Myanmar, Vietnam, Tuvalu and beyond, millions of people are facing and experiencing displacement as a result of climate change. DS has seen the human face of climate displacement up close in more than a dozen such countries, and our work to draw attention to the plight of the growing number of individuals, households or communities under threat has – out of necessity – expanded considerably since we began working on these issues in 2007.

Our work in the frontline states affected by climate displacement has revealed to us time and time again that research carried out by the Intergovernmental Panel on Climate Change (IPCC) reports, the Stern Review and many other studies on the extent to which climate change – including rising sea levels, heavier floods, more frequent and severe storms, drought and desertification – will cause large-scale population movements are often too future-focused, for climate displacement is *already happening today*. This novel form of displacement presents an urgent problem and challenging conundrum for affected communities, governments and the broader international community.

With a view to assisting these groups to better address the climate displacement dilemmas facing them, DS has dedicated most of 2011–2013 to building the foundations for a new normative framework to address climate displacement within States. Teams of DS experts spent countless hours reviewing the climate displacement literature; examining virtually all policy and legal documents dealing with climate displacement; travelling throughout the world to meet with government officials, academics, communities and those working in the field; and attending and presenting at climate displacement seminars and conferences. A solid six-month period was then spent drafting, re-drafting and re-drafting again and again (there were some 30 pre-final drafts of what became the new standard) and seeking expert inputs into the text. We placed an advanced text on the DS website and asked the public for comments, and many useful contributions were received from people from all corners of the planet.

Then, in mid-August 2013, representatives from Australia, New Zealand, Bangladesh, the Netherlands, Switzerland, the United Kingdom, Germany, Egypt, Tunisia and the United States came together in Red Hill, Victoria, Australia, and shared their backgrounds and expertise in international law, human rights and refugee law, forced migration, environmental change and United Nations policy creation to strengthen, stand behind and approve what became the Peninsula Principles on Climate Displacement within States (the Principles), which we believe is the first formal policy document of its kind in the world.

The Principles provide a comprehensive normative framework, based on principles of international law, human rights obligations and good practice, within which the rights of climate-displaced persons within States can be addressed. The Principles set out protection and assistance provisions, consistent with the UN Guiding Principles on Internal Displacement (upon which they build and contextualise), to be applied to climate-displaced persons.

The foundations of the Principles include the following:

- While climate displacement can involve both internal and cross-border displacement, most displacement will likely occur within State borders.
- Climate-displaced persons have a right to remain in their homes and retain connections to the land on which they live for as long as possible.
- Those who may be displaced have a right to move safely and to relocate within their national borders over time.
- Climate displacement, if not properly planned for and managed, may give rise to tensions and instability within States.
- Because climate change is a global problem, States should (upon request by affected States) provide adequate and appropriate support for mitigation, adaptation, relocation and protection measures, and provide assistance to climate-displaced persons.
- The international community has humanitarian, social, cultural, financial and security interests in addressing the problem of climate displacement in a timely, coordinated and targeted manner.
- There has been no significant coordinated response by States to address climate displacement, whether temporary or permanent in nature.
- Neither the United Nations Framework Convention on Climate Change (UNFCCC) nor its Kyoto Protocol either contemplates or addresses the issue of climate displacement.
- There is a need for a globally applicable normative framework to provide a coherent and principled approach for the collaborative provision of pre-emptive assistance to those who may be displaced by the effects of climate change, as well as remedial assistance to those who have been so displaced, and legal protections for both.

The Principles are divided into four operative parts: (a) general obligations; (b) climate displacement preparation and planning; (c) displacement; and (d) post-displacement and return.

General obligations include those pertaining to the prevention and avoidance of conditions that might lead to climate displacement; provision of adaptation assistance and protection measures; national implementation measures; and international cooperation and assistance.

Climate displacement preparation and planning includes climate displacement risk management; participation by and consent from affected individuals, households and communities regarding such preparation and planning; land identification, habitability and use; development of laws and policies for loss suffered and damage incurred in the context of climate displacement; and development and strengthening of institutional frameworks to support and facilitate the provision of assistance and protection.

Displacement requires state-based assistance to those climate-displaced persons who have not been relocated, including housing and livelihood matters and remedies and compensation.

Post-displacement and return sets out a framework for the process of return in the event that displacement is temporary and return to homes, lands or places of habitual residence is possible.

The Principles can now be practically applied in efforts designed to improve the prospects for climate-displaced persons, households and communities. They set out a framework for the collaborative provision of pre-emptive adaptation assistance, preparation and planning – and, if necessary, relocation, together with post-displacement matters and possible return to homes – *before* islands and coastlines are under water, *before* global warming worsens, and *before* glaciers melt and retreat even further than they already have.

Governments, international organisations and threatened communities can begin today to apply the Peninsula Principles to concrete situations where people are already facing or experiencing climate displacement. Importantly, the Principles take the correct view that communities are expected to play a fundamental role in organising themselves and outlining their future needs with regard to any looming – or ever-present – climate displacement threat. Communities need to organise themselves, come forward with their claims, and outline what the corresponding obligations of States are, based on existing human rights laws, to protect and respect the rights of those affected by climate displacement.

We know with increasing precision where climate displacement is already taking place or will take place, which people and how many are likely to be affected, and at least some of the – often land-based – solutions required to prevent and repair climate displacement. Thus, we now find ourselves at a juncture between theory and reality, between what could be and what clearly is. The Principles provide everyone concerned about the rights of climate-displaced persons, households and communities with a clear and consistent soft law basis for the practical actions required of us.

We, therefore, call upon all international agencies, governments (both national and local), communities, climate justice advocates and ordinary citizens to look carefully at the prospect of climate displacement wherever you live or work and try to apply the Principles as part of an effective strategy to treat climate-displaced persons, households and communities as the rights-holders that they so clearly are. Let us all work together towards this realistic and worthy objective, for together we can protect the rights of climate-displaced persons and resolve climate displacement the world over.

Scott Leckie and Chris Huggins

Note

1 www.displacementsolutions.org.

Acknowledgements

The process leading up to the development of the Peninsula Principles has been a fascinating one made possible by the collective efforts of scores of people across the planet who all care passionately about the displacement that is already being generated because of climate change. All of us who work on this issue in the field have witnessed climate displacement in its rawest forms: the embarrassed smile of a father unable to provide a safe home for his family as sea levels rise around them; the glint of hope in the eyes of children who know something is wrong, but precisely what it is still baffles them, even as it becomes ever more clear to those adults around them; and the worrying looks of gradual despair we have seen in the eyes of well-meaning, honest and caring politicians (yes, they still exist in the world!) as they contemplate the immense challenges facing them in the era of climate change.

Displacement Solutions has been engaged in the issue of climate displacement since our founding in 2006 and since then has continually expanded our attention to this problem in an ever-growing array of countries in virtually every corner of the world. We have tried to tackle these challenges head on and have consistently taken a rights-based and solutions-based approach towards this expanding crisis, whether in Bangladesh or Tuvalu, or Alaska, Panama or Myanmar and beyond. The more we worked on these issues, the more it became clear that besides the problem-solving work we engaged in, the research, the meetings with officials and communities and countless visits to field locations in numerous countries, the idea of a new global rights framework on the question of climate displacement would be an endeavour worth pursuing, notwithstanding how difficult it might be, and one that would assist climate-displaced communities and their governments to find viable ways to address what had previously often been seen as a problem so large that it was effectively without solution.

As our awareness of the problem grew and as our work expanded, I received a synchronistic phone call from Professor David Hodgkinson who had embarked on an effort to achieve the adoption by the international community of a new convention elaborating the rights of climate-displaced people. He had called to see if I would be interested in joining forces. After a very lengthy and infinitely pleasant conversation, I let David know that I would certainly be on board, but that there would be no way that I could dedicate the years in the UN headquarters in Geneva needed to achieve the finalisation of such an ambitious new multilateral treaty. After having run myself ragged at the UN between 1988 and 2005, drafting and then pursuing the approval of scores of new legal texts, after becoming a father and moving to far-off coastal Australia, I was out of steam for such things, and had already decided that I would leave such work to the newcomers to the human rights scene. As time went on, I continued to offer strategic advice on the convention to David and his group even as it became increasingly clear that vital political support required for such an effort was simply not forthcoming. Similarly, while several UN agencies clearly sought to expand their mandates, albeit only marginally, to address the displacement consequences of climate change, it also became clear that most donor States were simply not interested in such expanded mandates. Nevertheless, some of the more reasonable States, particularly Switzerland and Norway, clearly wanted to do something, and with those sentiments decisions were made to establish what became the Nansen Initiative which was put into place to explore the development of a protection agenda to deal more effectively with the gaps in international law on cross-border movements of people due to disasters, climate change and other environmental factors. Since 2012, the NI has carried out some excellent work in generating greater understanding of these types of cross-border movements. It concluded its important work in 2015.

All of these factors taken together, combined with our growing climate displacement work in the field, then, led to the establishment by DS of our Climate Displacement Law Project (CDLP) in 2012. This formed part of the DS Climate Change and Displacement Initiative which has been in place since 2007 when our solution-oriented approaches to this question began in earnest. The CDLP thus became a vehicle by which we could seek to bring the best and brightest climate displacement minds, UN officials, judges, lawyers, jurists, NGO leaders and community-level practitioners among them, together in a common effort to develop a new normative framework, driven from the bottom up. We would consult the world at large, visit every climate displacement hotspot across the globe, read every article, book and analysis ever written on climate displacement, and gauge existing policy, law and practice the world over to determine how best to approach the growing problem of people losing their homes and lands because of climate change. And this is precisely what we did throughout most of 2011, 2012 and 2013, a 1,000-day process that eventually resulted in the text that this book is all about, the Peninsula Principles on Climate Displacement within States.

As I noted above, this process involved countless people in places too many to mention here, but one person above all others made this process possible, and that is Mr Andre Hoffmann whose private Foundation funded the majority of the budget of the project. Andre and his assistant Sylvie Leget have been extraordinarily supportive throughout the entire process and everyone who believes in the Peninsula Principles and sees their value has Andre and Sylvie to thank, for without them it is unlikely this project would have ever been able to achieve what it has been able to thus far. We are also grateful, of course, to the other donors who have helped this process come to fruition, and here we would also very much like to thank the German Ministry for Development Cooperation (BMZ) for their support for our work on land solutions to climate displacement which played a vital role in the development of the Principles. We are particularly grateful for the detailed comments given by the German government to early drafts of the Principles, which helped inform the text in very useful ways. We are also extremely grateful to the Pictet Foundation, the government of Liechtenstein, the International Federation of Red Cross and Red Crescent Societies, and other donors who continue to make our climate displacement work possible.

While DS experts spoke with thousands of people across the globe in the lead up to the drafting process and are grateful to all of them for their views and wisdom, it is really David Hodgkinson who deserves to be singled out for his efforts in formulating many of the ideas which eventually became aspects of the Principles. Along with David, the extremely hard work of those who actually attended the final drafting sessions in August 2013 is deeply, deeply appreciated. Many of those who assisted in this regard prepared chapters for the present volume and for that, as well, we are very grateful. Along with them, Rick Towle of the UN High Commission for Refugees, Arifur Rahman of Young Power in Social Action in Bangladesh, Dutch law professor Ingrid Boas, Tamer Afifi from the United Nations University in Bonn and others all made vital contributions.

The many fine folk at EcoHummingbird Retreat where the final drafting meetings were held went to extraordinary lengths in creating a setting where everyone was made as inspired and comfortable as possible for the important work that was to be done there. The rainforest in which the Retreat is located served as an invaluable source of pristine beauty and rejuvenation for many of us, and the frequent visits from king parrots, kookaburras, cockatoos, gallahs, lorikeets and so many other colourful birds that Australia is blessed with kept us all going as they flew from ancient blue gum to wattle to acacia.

Special thanks to Jon Staley and Haydn from YouthWorx who spent three days with us filming the entire process and who produced an excellent short film detailing the process leading up to the adoption of the Principles. Graphic designer extraordinaire Craig Brown of Arteria Design also deserves hearty thanks for his stellar efforts in getting a designed version of the Principles ready within hours of their adoption and generally for his much appreciated work for Displacement Solutions in so many invaluable ways.

As always, it was again a great pleasure working with Chris Huggins to bring this book to fruition. Chris and I have now worked together on housing, land and property issues across the globe and written and edited several books and reports along the way, and if all work mates were like him, the world would surely be a better place.

Finally, the most profound thanks of all go to my little family in our little house on our little peninsula with our little dog, all of which proves again and again that with infinite love the possibilities are truly infinite.

Scott Leckie
Blairgowrie, Australia
March 2015

1 Using human rights to resolve the climate displacement problem

The promise of the Peninsula Principles

Scott Leckie

As the consequences of climate change become increasingly apparent along the coastlines and other vulnerable areas of a sadly expanding array of countries, a growing number of efforts to understand, address, legislate and ultimately resolve climate displacement have emerged. Many of these efforts, for example by Tulele Peisa in Papua New Guinea or the Guna indigenous group in Panama, have evolved at the most grassroots of levels among community members who are themselves staring climate displacement in the face. Others, such as Displacement Solutions which works globally or Young Power in Social Action (YPSA) in Chittagong, Bangladesh, work in the field at multiple levels; from the UN and governments to community groups and NGOs working to address climate displacement, through research, standard setting and media efforts, all have adopted a rights-based approach, grounded in particular in housing, land and property rights. Yet others such as the Nansen Initiative on Disaster-Induced Cross-Border Displacement or many of those working within academia focus primarily on the numerically small issue of cross-border movement of people; far more people will be internally displaced by the forces of climate than will flee across international borders. Some such actors operate from a perspective grounded in political pragmatism which often ignores or downplays rights-based, community-driven approaches to this question. These latter groups often use terms such as 'climate migration' to describe climate displacement, as if individualised choices to move house to better locations without governmental and rights-based protection interventions represent an acceptable response to an extremely complex, highly sensitive and rights-threatening process.

This book, and the themes and Principles it addresses, takes the view that people everywhere who are threatened with the loss of their homes and lands because of climate change factors beyond their control are holders of rights in their persons and in these homes and lands, and consequently need to be treated as rights holders by the relevant authorities. As holders of rights (focusing here on housing, land and property rights and their many close cousins: rights to freedom of movement, rights to privacy, right to respect for the home, right to be free from discrimination, etc.), people who flee their abodes because of the effects of climate change, whether acute or long term in nature, also, therefore, need to be able to access appropriate and enforceable restorative remedies. Such remedies would repair this displacement and restore the rights they are in the process of losing due to this unwanted and unsought displacement from the place they call home. They need new homes for lost homes, and new lands for lost lands.

Those of us who take rights-based approaches to these matters are neither content with, nor do we advocate for, policies that hinge only upon 'climate migration'. We recognise – based on our years of work with climate-affected communities throughout the world – that as a default policy migration alone is not enough: certainly some will be happy to migrate, but many more will suffer great detriment and misfortune in the process. Many stand to lose all they have, including their homes, their lands and their investments, if they are forced to flee rising seas and other manifestations of global warming; and they have every right to expect that their governments and the international community will establish firm policies, laws, institutions and programmes designed specifically to assist them to regain the rights they have lost in this hugely unfair and arduous process of climate displacement.

Population movements due to climate change are *already happening* across the globe. Tens of thousands of Bangladeshis have already given up on low-lying coastal areas and are moving to higher ground in the east of the country, or staking out claims in Dhaka's already crowded slums. Thousands of members of the Guna indigenous group in Panama are moving from their Caribbean island homes to the mainland as their traditional villages are slowly inundated. People in the Solomon Islands have begun evacuating long-inhabited settlements for safer areas. Inhabitants of Newtok village in the US state of Alaska have begun their trek to nearby Nelson Island because of severe erosion making their home village no longer habitable. The Carteret Islanders in Papua New Guinea have slowly begun their relocation to the larger island of Bougainville, while the Fijian government has identified 676 villages in need of relocation. Small island states such as Kiribati, the Maldives, the Marshall Islands and Tuvalu which rightfully receive the lion's share of media attention on the looming demise of their islands and atolls, of course, are already putting plans into place which ultimately envisage the entire evacuation of their beloved island homes.

These are the frontlines of climate displacement today and sadly are just the first of what will surely be an ever growing number of places now called home by people and communities, and sometimes entire nations, that will be forced to endure climate displacement. We now find ourselves at a global policy crossroads on how best to address the crisis of climate displacement, and we believe that the Peninsula Principles provide a consistent, coherent, manageable and practical rights-based normative framework that can be used today to find the best possible outcomes for those forced to leave their homes due to climate change. Migration and minimalist attitudes that back such approaches are wholly inadequate as a solution to this crisis.

Prevention and remedy

Those of us working daily on the climate displacement challenge with the actual people who are already facing this crisis experience the relatively novel situation of working with vulnerable communities to prepare them for changes to their residential realities which go to the very core of what the promise of human rights is all about. When they view the displacement being caused and that which will be caused by climate change as an issue of human rights, advocates, lawyers and activists find themselves increasingly placed in positions not always typically associated with human rights work. While human rights advocacy does indeed involve a range of preventive and precautionary protective measures, more often than not much of the energy invested by human rights advocates in the elusive quest for justice is spent following human rights abuses which have already occurred. When human rights efforts involve prevention, this normally implies helping people to stay put, not advocating their relocation.

Looking specifically at human rights lawyers who focus, for instance, on the enforcement of housing, land and property rights, most of their work hinges on either preventing and halting planned evictions before they are carried out, securing adequate housing for the homeless or ill-housed, or securing restitution rights for internally displaced people (IDPs) and refugees seeking to recover their original homes. In the age of climate change, however, efforts to address the human rights dimensions of climate displacement are based primarily on preventing the negative human rights consequences of what many world leaders now recognise as the single largest problem facing the global population: the crisis of climate change.

The main pre-emptive efforts of preventive human rights work to address looming climate displacement of course seek to shape adaptation measures in such a way that they assist in preventing more dramatic outcomes such as migration and displacement. Where adaptation fails or is inadequate (as it increasingly is), the focus then turns to the option of coordinated, planned, land-based and adequately resourced relocation, largely domestic in nature, from affected areas to safer land where life can begin again for the world's growing climate-displaced population. Indeed, not only is climate displacement under way, but so too is planned relocation in far more places and affecting far more people than is commonly known. Many of them are mentioned in this volume.

With the emerging recognition that some form of pre-emptive relocation will need to occur to protect threatened communities, growing numbers of analysts, scholars, civil society groups and governments are devoting attention to the challenges of climate displacement.[1] As noted, some of this attention has focused primarily on the question of migration and how migration pathways can be opened for those forced to flee their homes due to environmental factors, including climate change. Others conversely recognise the importance of migration as one of many policy options open to climate-displaced persons, but place a far firmer, rights-based emphasis on the issue of planned and voluntary relocation, based on the belief that only through guided, properly planned and financed relocation can the full spectrum of human rights of those affected be respected, protected and fulfilled. Simply supporting the contention that migration alone will be enough is wholly insufficient.

Moreover, given the prevailing absence of any form of compensation or financial assistance for those affected in virtually all of the communities where climate displacement is already under way, and bearing in mind that most people in most places have most of their assets linked to their homes and lands (the precise areas most affected by rising sea levels, coastal erosion and other climate change consequences), if people are left simply to migrate on their own, they will likely face both financial and human rights losses. These are not fair, nor are they circumstances upon which democratic societies based on the rule of law and justice can be built, let alone sustained. Given the rather frenzied and sudden interest by insurance companies in climate displacement, we can be certain that few will receive the insurance protection they rightly deserve and rights-based approaches will need to fill the gap.[2]

Considering the sheer scale of current and future climate displacement, it is clear that managed approaches will be difficult, risky and expensive, but ultimately they will be unavoidable if rights are to be protected in full. Governments will need to establish specific institutions to guide these processes, allocate public and private funds towards solutions, and ensure that everyone affected, everywhere, knows exactly where to turn in the event they are forced to leave their homes and lands. Once again, migration is shown for what it truly is: the easiest way out for governments unwilling or unable to more proactively support the rights of climate-displaced people. As an approach to climate displacement, migration by itself is simply not enough.

There are a wide spectrum of ways by which these various processes can be built, and normative frameworks upon which they can be based, one of which is the theme around which this book is constructed, namely the Peninsula Principles on Climate Displacement within States. The Peninsula Principles are the result of a two-year process following more than a decade of field experience with climate-affected communities, involving people and institutions from across the world that led to these 18 universally applicable Principles that can form the basis for effective domestic law- and policy-making designed specifically to secure the rights (in particular the housing, land and property rights) of everyone affected by climate change.

Those of us who were involved in the process leading to the final text of the Principles intentionally excluded reference to the question of cross-border displacement, believing that this was the domain of the Nansen Initiative. Far more important than this gesture, though, is that the ultimate purpose of our efforts is to assist national governments to solve the climate displacement challenges they face internally in the most rights-based manner possible. Because so many more people will remain within their countries than will attempt to leave them, it was more than apparent that a solutions-based approach to the question was infinitely more valuable than one based primarily on theory, law or ideology, as so many of the efforts to address climate displacement seem to be.

The process leading up to what became the Principles emerged after it became increasingly clear that the political support required for the adoption of any of the several proposed new Conventions on the Rights of Climate-Displaced Persons was – despite the efforts of many people and organisations to generate this support – far from forthcoming. Following discussions between the author and David Hodgkinson of the University of Western Australia, and financial support by a Swiss philanthropist for the Displacement Solutions' Climate Displacement Law Project, the process began. Much can be said about this process, but perhaps the most important point to be made at this stage is simply that the Principles are the result of years of global discussions, legal analysis, field research and missions to virtually every climate displacement hotspot throughout the world. Beyond the simple but all too often overlooked premise, that everyone forced to depart their homes and lands because of the consequences of climate change should have an *enforceable right* to acquire a new place to live, consistent in every way with their full spectrum of housing, land and property (HLP) rights, the Peninsula Principles are founded on the clear and observable truth that the vast majority of climate displacement will occur internally within the borders of States, and thus the governments of whatever country is affected in this way will retain the primary responsibility for protecting – using the terminology of the Principles – these 'individuals, households and communities' and the rights they are intended to enjoy under domestic, regional and international legal regimes.

The causation conundrum: what conundrum?

Bring up the topic of climate displacement in a discussion, particularly among the rather conservative voices of some lawyers and academics involved in the field, and sooner rather than later the question inevitably emerges of what could be called the 'climate displacement causation conundrum': namely, how does one prove that it was climate change that caused the displacement in question, and how do we ensure that those presenting themselves as persons displaced by these causes are legitimate claimants? An almost obsessive concern with this issue has become apparent within certain legalistic sections of the climate-displacement community. These highly reductionist views ultimately benefit governmental agendas that seek to do as little as possible on this issue, thus forcing climate displaced people into an even more difficult position, leading only in one direction: migration. Focusing on migration, of course, is not a rights-based approach to what should be seen and treated as an inherent human rights issue.

Simply put, those who choose to place a primary emphasis on the question of causation, rather than on resolving the losses of rights and the actual de facto circumstances facing those who are forced to move due to climate change factors beyond their control (and for which they bear absolutely no responsibility whatsoever), do little to serve the millions of people already facing the loss of their homes and lands. Such voices often argue that those seeking protection will need to convincingly prove that it was climate change, and climate change alone, that was the cause of their displacement. Failing to prove this to the relevant authorities, therefore, would result in affected people and communities ending up with no legal basis for claiming protection from their State or another State in which they may find themselves: a tragic outcome to an already tragic situation.

With all due respect to those advocating such views, the real-life situation on the ground in communities that are already enduring the effects of climate displacement, as well as those who are contemplating future displacement as global warming worsens and sea levels continue to rise, is that households are rarely – *if ever!* – forced into a position of having to prove that climate change was the exclusive cause of the displacement they are enduring or will endure. In all of our work with groups in Bangladesh, Kiribati, the Maldives, Panama, Papua New Guinea, the Solomon Islands, Tuvalu, the US and elsewhere where various types of climate displacement and relocation are already under way, we have *never* encountered either arguments by government officials or particular evidence or proof-based crises that have prevented affected populations from being recognised as legitimate claimants of new homes and lands.

While the causation issue may be of relevance in the comparatively rare instances of individual asylum seekers arguing that climate change was the cause of their alleged persecution, for the vulnerable coastal and inland communities of the countries just mentioned, the causation issue is in practice much ado about very little. While scientists and lawyers may bicker about issues of causation, the citizens and local governments already facing climate displacement are far beyond this point of analysis. Institutions including the various local governments and national government in Bangladesh, the government of the Autonomous Region of Bougainville, the Fijian government, the national government of Panama and many others have made decisions to relocate threatened populations to improved and safer locations, in a manner consistent with the wishes of the people and the organisations supporting them, without ever having to prove scientifically that the cause of the displacement in question was climate change and climate change alone. Such local and national governments simply make such decisions on the basis of facts on the ground. Responsible governments in the real world evidently are not that concerned as to whether or not a community which is no longer able to live a full and dignified life in a particular location is threatened by climate change or some other combination of environmental causes, whether acute or longer term in nature.

What matters, in fact, is that communities themselves are actively asserting their own rights, organising themselves to find better solutions to their plight, and approaching governments for assistance in achieving better outcomes. Governments, of course, don't always provide the best possible results, nor do they always sufficiently prioritise the actions required to find rights-based solutions. History shows clearly enough that the people themselves as rights-holders need to be central players in any attempt to ensure community-level improvement.

In debating circles the unfortunate over-emphasis on causation issues is a fortunately rapidly decomposing red herring comprising arguments in support of a position which may sound reasonable during theoretical discussions by those safely ensconced in ivy-covered buildings, but which simply bears little or no resemblance to real-life circumstances affecting ever-larger numbers of people. Clearly what is infinitely more important than any preoccupation with causation is determining who is under threat from the effects of climate change, where they reside, and the degree of urgency associated with finding rights-based solutions to their predicaments.

Who is a climate-displaced person?

In a manner similar to the approach taken on the causation question, we need also to consider who is to be classified as what the Principles call a 'climate displaced person', and consequently who can expect to be protected by relevant laws and normative frameworks. Of all the individual principles negotiated during the drafting process that led up to the final text of the Principles, Principle 2 on definitions clearly required the most effort and creative thinking. Principle 2, of course, reads as follows:

For the purposes of these Peninsula Principles:

a 'Climate change' means the alteration in the composition of the global atmosphere that is in addition to natural variability over comparable time periods (as defined by the IPCC [Intergovernmental Panel on Climate Change]).

b 'Climate displacement' means the movement of people within a
State due to the effects of climate change, including sudden and
slow-onset environmental events and processes, occurring either
alone or in combination with other factors.

c 'Climate displaced persons' means individuals, households or communities who are facing or experiencing climate displacement.

d 'Relocation' means the voluntary, planned and coordinated movement of climate displaced persons within States to suitable locations, away from risk-prone areas, where they can enjoy the full
spectrum of rights including housing, land and property rights and
all other livelihood and related rights.

The drafters of the Principles devoted a large amount of energy and time
to developing what ultimately became quite clear-cut, inherently simple yet
sophisticated ways of addressing both the issues of causation and determining who should be classified as a climate-displaced person in a manner
that would assist affected people, governments and the lawyers and judges
who may eventually become engaged in these efforts. Recalling that the
process of developing this terminology involved inputs by hundreds of
experts and ordinary citizens from across the globe, as well as judges,
lawyers, legal scholars, UN officials, law professors and activists who were
all present during the final drafting phase of the process, we feel quite con-
fident that the language developed in Principle 2, in particular Principles
2b and 2c, rather eloquently elaborate concise and actionable definitions
of both 'climate displacement' and 'climate displaced person'. Principle 2b
provides an all-encompassing approach in its definition of climate displacement and Principle 2c includes three key points relating to the definition of
'climate displaced persons' or CDPs.

First, the definition takes a wholly inclusive approach that covers 'individuals', 'households' and larger 'communities', in other words, everyone
who could possibly be affected by climate displacement. This terminology
is used throughout the entire body of the Principles. No one, therefore,
should ever fall between the cracks which could, otherwise, create a world
where some would be classified as CDPs and others not.

Second, Principle 2c speaks of those CDPs who are 'facing' climate displacement. This intentionally general term was selected after very extensive
negotiations to reflect future climate displacement based on climate modelling and other projections, while at the same time being flexible enough to
also include reference to those who may not as yet have been forced to flee
their homes, but whose movement is effectively imminent. Some may argue
that such language is perhaps too inclusive or conversely too imprecise;
however, there is nothing in the term 'facing' that would prohibit a judge
from making an appropriate determination about the circumstances facing
a threatened CDP or group of CDPs, or to stop CDPs themselves from successfully claiming protection for the climate displacement that they face,

whether now or in the future. It is a sufficiently precise term to distinguish clearly between real CDPs who face climate displacement and others who may claim to be CDPs because of the potential benefits this may accrue for them. It is a term that can easily be tested in real-life scenarios in coastal areas, rural areas, urban areas and beyond, and one that can equally generate an ever-growing web of interpretation and even jurisprudence as to what this term means in practice.

Finally, and third, Principle 2c speaks of CDPs who are 'experiencing' climate displacement. This refers, of course, to CDPs who are currently enduring climate displacement. It would address groups such as the Carteret Islanders in Papua New Guinea, those dwellers in the Solomon Islands who are now moving from locales such as Lau Lagoon and the atoll of Ontang Java, residents of 676 villages who have been scheduled for relocation in Fiji, the 30,000-strong Guna indigenous island dwellers who are in the process of moving to safer land on the Panamanian coastline and all others across the world who are already dealing with the human tragedy of climate displacement, but who are yet to receive the protection and respect that human rights laws and the Peninsula Principles are meant to bestow upon them.

As a climate-displaced person, on whose door do I knock?

Beyond these definitional issues, the Principles are also designed to encourage governments everywhere to develop the institutional frameworks required to secure the rights of all CDPs. Without such institutions, there is little hope that a rights-based approach to climate displacement will take effect in the manner required. Part of the motivation for emphasising such institutional frameworks stems from our work in a growing number of countries already grappling with climate displacement, where we have almost universally found the needed institutional mechanisms to be sorely lacking. One compelling question has repeatedly presented itself during work with CDPs, which can be paraphrased simply as: 'As a climate-displaced person, on whose door do I knock?'

Whether in Kiribati, Bangladesh or Panama – and all the other countries in the frontlines of climate displacement – in no instance have we yet been able to identify the precise government agency or people within government who have been specifically entrusted with official competencies to secure the rights of climate-displaced persons. Yes, in Kiribati people can – as they do – call the President's Office in Kiribati and ask for more sandbags for a dilapidating seawall or perhaps even flag him down in his car on the one road that traverses the long and narrow capital of South Tarawa, or they may pay a visit to the Kiribati Adaptation Project for assistance, but there is certainly no specific office or official in place to deal comprehensively with the spectre, both real and looming, of the climate displacement process itself. In response to a similar state of affairs in Bangladesh, as a core element of the DS-YPSA Bangladesh HLP Initiative, DS supported and carried out an in-depth research project which resulted in the first ever attempt to systematically determine every government entity in one country which maintains, in one way or another, any legal or political competencies to resolve climate displacement.[3] What we found in Bangladesh is simply that literally dozens of different ministries, agencies and programmes have some measure of official competence when broader climate change and land issues are addressed, but no single agency can be rightly classified as *the* agency entrusted with the important task of protecting the full spectrum of rights held by those displaced due to climate change.

In all climate-affected countries there are indeed climate change focal points within various ministries and disaster prevention agencies, and in some, particular offices have been established to manage adaptation funding resources. Adaptation efforts may very occasionally have programmes specifically addressing the displacement dimensions of climate change. However, nowhere have we yet identified a stand-alone agency with official competencies to coordinate all aspects of the resolution of climate displacement, despite the scale and very real nature of the problem.

Linking this situation back to the Peninsula Principles, the normative framework that the Principles outline provides an ideal manner by which climate displacement policy at the domestic level can be developed, and can also provide the basic normative foundations for the establishment of national- and local-level agencies entrusted with systematically resolving the climate displacement challenge. Just as ministries, departments or agencies are routinely established when new governments with new priorities wish to enact their desired policies and projects, so too can any government, and preferably all of them, establish new state-based agencies, or even ministries, to coordinate solutions to climate displacement. Having a focal point such as an Office on Climate Displacement and Relocation in place within the heavily affected countries would signify a huge step forward in the global quest to protect the rights of CDPs wherever they may be.

The basic choice: bigger slums or better communities?

The Peninsula Principles are important also for the role they can play in helping communities and governments to find viable and rights-reinforcing managed solutions to climate displacement. In many of the countries most threatened by climate displacement, in the absence of adequately financed, planned and politically backed policies and laws actively supporting climate-displaced persons and communities, the one and only option that will be available to ever growing numbers of CDPs will be a new home in an already overcrowded urban slum. This need not be the case, and diligently applying the Principles can assist in giving affected communities the types of solutions to their displacement that human rights laws envisage for them.

A hands-off, migration-only approach to climate displacement – which is so far the norm – will, with certainty, lead to a growing number of people facing the loss of their rights and a subsequent growth in slums and social and economic impoverishment, outcomes that no one could consider desirable. Despite the rising voices of those proclaiming the virtues of migration, we can rest assured that in the absence of well-thought-out government plans and policies to address the growing problem of climate displacement, most of those affected will be migrating not to better homes on better lands in better locations, but rather to drab and often dangerous neighbourhoods which are in all likelihood far worse than the areas whence they came.

Is this an acceptable approach to climate displacement? Is this the outcome we want to see in the coming years and decades as climate displacement grows and grows, as it is guaranteed to do? Can we accept that these rights-holders are simply cursed by bad luck and misfortune and left to fend for themselves in a world absent of any redress for these human rights losses, where compassion has gone completely missing?

The crucial climate change and HLP link: land, land, land

It may come as a surprise to some, even those deeply engaged in the climate change movement, that large numbers of people and communities are *already* being forced to move because of climate change, and even more so, that land in some instances is already being sought and acquired for the relocation of climate-displaced people and communities.[4] At its core, this is the approach taken by the Peninsula Principles: individuals, households and communities facing or experiencing climate displacement have every right to expect that their governments and the broader international community will *actively* support measures, often involving the acquisition and allocation of land, as a central means of protecting the HLP rights of climate-displaced persons wherever they may reside. Again, migration as policy will simply fall far short of the mark.

Several earlier publications have specifically addressed the land solutions approach to climate displacement, and have attempted to make the point that land lies at the core of protective strategies designed to secure the rights of those affected.[5] Our global land needs models suggest that in total anywhere between 12.5 million (roughly the size of Tasmania) and 50 million acres (the approximate size of Uganda) of land would be required to provide various land-based solutions to the world's climate-displaced population. This may sound like a lot, but in fact it represents a minuscule 0.03–0.14 per cent of the Earth's land surface as it exists today; amounts of land certainly within our collective grasp should we pursue such strategies to resolve climate displacement.

Some countries have begun to acknowledge the need to secure land parcels for people displaced by climate change, but obstacles remain.

We have developed plans to assist CDPs to secure land parcels in Bangladesh, Panama and Papua New Guinea, and are researching similar actions in Kiribati, the Maldives, the Solomon Islands, Tuvalu and elsewhere. National-level land estimates are beginning to be made by climate activists, and in one recent case Bangladeshi researchers estimated that some 1.75 million acres of non-agricultural land would be required to rehabilitate the climate-displaced population of that country. Efforts by DS and YPSA through the joint Bangladesh Housing, Land and Property Rights Initiative have identified ten land parcels in Chittagong district in the eastern part of the country which can serve as the first parcels to be acquired and allocated, as part of a climate displacement resolution programme, for roughly 1,000 climate-displaced households, with the first parcels expected to be available for relocation in 2016. It is hoped that this will create the successful precedents needed for this approach to become national policy in Bangladesh in the coming years.

As is widely known, movement is under way from the Carteret Islands in Papua New Guinea to the larger island of Bougainville, which lies roughly 100 km to the south. Thus far, several dozen Carteret Islanders have relocated to an area on Bougainville known as Tinputz, a very challenging and difficult process that was captured in graphic form in the Academy Award-nominated film *Sun Come Up*. Efforts have been under way for several years to acquire land on Bougainville by the group Tulele Peisa and others, but this process has proved far more complex and stupefying than was originally expected. In 2008, DS put together a detailed plan which – had it been implemented by the government at the time – would have resulted in more than 2,700 hectares of land on Bougainville being reserved for the possible resettlement of the entire population of the Carteret Islands. The funds needed were made available, local groups took charge of the relocation and it looked as if a winning formula was solidly in place. Then as quickly as the government funds were allocated, the money was suddenly nowhere to be found and as a result the relocation plans were shelved.

In Panama, some 30,000 members of the Guna indigenous community in the Gunayala region of Panama, where Guna indigenous communities live on approximately 40 islands along the eastern Caribbean coastline, will ultimately be displaced as a result of climate-related events. A 2014 report by Displacement Solutions, *The Peninsula Principles in Action: Climate Change and Displacement in the Autonomous Region of Gunayala, Panama*, highlights the impacts of a series of disasters and weather-related events on the Gunayala islands over the last ten years and concludes that some 30,000 Guna people will need to be relocated in the coming years. So far, no official planning has been undertaken to address the situation and there is nowhere with sufficient housing or infrastructure for them to go. Panama has established a very impressive disaster risk management system which, however, has been slow to recognise the risks presented by climate change displacement. This is the case in many other parts of the world where climate change is already taking place. In the absence of government support, the Guna people have already initiated their own efforts to start the process of relocating to the mainland before the islands become uninhabitable. For almost four years the people of the island of Gardi Sugdub have been working to relocate their community of some 1,000 inhabitants to a safer place on the mainland. In 2010, the community on Gardi Sugdub decided to relocate to the mainland and created a Comisión de la Barriada or 'neighbourhood commission' to organise the relocation process. By mid-2014, the date of the fact-finding mission's visit to Gunayala, a total of 300 families (+/− 1,500 people) from Gardi Sugdub had signed up to be relocated to the mainland. Approximately 200 of these families are currently living on Gardi Sugdub, with the other 100 families originally from the island now living in Panama City. Despite the fact that four years have passed since the community decided to move, the relocation has yet to take place. After the initial decision was taken to relocate in 2010, the community acquired the necessary land on the mainland to commence the process. Seventeen hectares of land was donated by several families from the community for the site where the first houses would be built. Arrangements to clear the site were put in place, as land located on mainland parts of Gunayala is covered by one of the most dense and well-preserved forests in Panama. The committee also approached different governmental agencies requesting support, and secured agreement from the Ministry of Housing to implement a project to build the first 65 houses at the relocation site. A planned project to provide alternative housing for a small number of those forced to relocate was later abandoned by the Ministry of Housing.

The monumental relocation of the 321 coastal-dwelling people of Newtok Village on the western coast of Alaska in the United States to nearby Nelson Island is the start of what is likely to be a far larger relocation process of indigenous Alaskans as their long-inhabited coastlines become increasingly unviable.[6] Climate displacement is also occurring in various locations in the Solomon Islands including the well-known case of Lau Lagoon, and Ontang Java atoll, where potentially the entire population is to be relocated. Customary law in the Solomons has played an instrumental role in assisting those who need to relocate to find new land resources within the customary land area of their *wantok* or cultural grouping. Villagers from almost 700 locations in Fiji will need to find new land in coming decades. The government of Kiribati recently purchased 6,000 acres of land in neighbouring Fiji as a hedge against possible resettlement if seas rise too quickly, but thus far has received no assurances from Fiji that the people of Kiribati might be allowed to migrate there. This is just the beginning.

It is becoming increasingly clear precisely where the effects of climate change are going to be most intensely felt. In all of the countries where we have worked directly on developing rights-based strategies to resolve climate displacement, however, the laws, policies and plans required to find sustainable solutions to climate displacement that protect locals' rights are still not yet in place. Many states thus far have been reluctant to proactively find land to house displaced people, and this needs to change. The time has come to start a meaningful global discussion on how climate-displaced people can be helped to address the personal consequences of land and property loss, flight from places of habitual residence and the need to find new homes and lands for displaced people. Using the Peninsula Principles can help.

The world clearly has more than enough available land to provide for the needs of those facing and living with climate displacement, even if we base those needs on worst-case scenarios of sea level rise. We need to explore in far more depth the particular role of land as a solution to climate displacement, and how land can be acquired for climate-displaced communities. A community-led, land-based approach to resolving climate displacement holds out considerable prospects for achieving the dual aims of protecting the rights of climate-displaced people, while simultaneously securing for them sustainable, viable and practical new places of residence. It will not be easy, but climate displacement *can* be fixed, and land is sure to figure prominently in the quest for solutions to this growing crisis.

The Peninsula Principles provide a way forward for governments, UN agencies, NGOs and, most importantly, communities themselves, to guide the contents of the laws and policies needed to protect the rights of these increasingly vulnerable groups. There is little chance that the world's governments will be able to stop climate change and the effects it will have across the world. But starting today, and using the Principles as a source of inspiration and guidance, efforts can begin that yield not despair, landlessness, poverty and conflict, but rather the enforcement of rights, the building of new communities and solutions to a global crisis that has only recently begun to show its more heinous face.

Notes

1 See, for instance, Scott Leckie (ed.) *Land Solutions for Climate Displacement*, Abingdon: Routledge/Taylor & Francis, 2013; Scott Leckie, *Finding Land Solutions to Climate Displacement: A Challenge Like Few Others*, Geneva: Displacement Solutions, 2013; UNHCR, *Planned Relocation, Disasters and Climate Change: Consolidating Good Practices and Preparing for the Future: Report* (Sanremo, Italy, 12–14 March 2014), Geneva: UNHCR, 2014.
2 Naomi Klein, *This Changes Everything: Capitalism vs The Climate*, London: Allen Lane, 2014.
3 Displacement Solutions (DS) and Young Power in Social Action (YPSA), *Bangladesh Housing, Land and Property (HLP) Rights Initiative: Climate Displacement in Bangladesh: Stakeholders, Laws and Policies – Mapping the Existing Institutional Framework*, Geneva: Displacement Solutions, 2014.

4 Displacement Solutions, together with the UN Environment Programme and the *New York Times* have worked with photo-journalist Kadir van Lohuizen since 2011 documenting climate displacement in more than a dozen countries. The photographs, interviews and films have been assembled in the exhibition *Where Will We Go* which was launched at the 20th Conference of the Parties to the UN Framework Convention on Climate Change in Lima in December 2014. The exhibition is now travelling the world and is on permanent display in various venues.

5 See e.g. Leckie (ed.) *Land Solutions for Climate Displacement*, and *Finding Land Solutions to Climate Displacement*.

6 For an excellent and detailed overview of this process, see: Robin Bronen, 'Climate-Induced Community Relocations: Creating an Adaptive Governance Framework Based in Human Rights Doctrine', *NYU Review of Law and Social Change*, 35: 357.

2 A rights-based approach to climate displacement

Khaled Hassine

The climate displacement anastomosis

Anastomosis designates in medical sciences the connection of separate parts, which may occur naturally or during an embryonic development, surgery, trauma or by pathological means.[1] By way of analogy, in climate displacement terms the present-day reality in which the artificial distinction between the already virulent displacement of people in search of safety from the scourge of severe or permanent environmental change and climate change has finally given way to an appreciation of the intrinsic nexus between the two dimensions.

In its first Assessment Report, dating back to 1990, the Intergovernmental Panel on Climate Change[2] (IPCC) had already ascertained that the greatest single impact of climate change will be human migration. Projections and estimates of numbers of people who will be forced to move as an immediate result of climate change remain controversial and diverge depending on the specific circumstances and the parameters used.[3] For instance, the IPCC in its report estimated that by 2050, 150 million people could be displaced by climate change-induced phenomena such as desertification, water scarcity, floods and storms,[4] whereas a more recent and frequently cited estimate is that 200 million will be forcibly displaced by the year 2050, losing their home, land and property.[5]

While the linkages between climate change and displacement are complex and cannot entirely be predicted,[6] the views of scientists have by now converged to agree that the situation has become irreversible[7] and that in addition to mitigation, designed to limit the phenomenon of climate change and its impact as far as possible, suitable adaptation strategies need to be devised. Climate displacement will be one of the many ways in which affected populations will adapt to their changed environment, and human mobility may reach an unprecedented order of magnitude, raising significant challenges.[8] However, it was only in 2010, with the Cancún Adaptation Framework, that migration, displacement and planned relocation were recognized as forms of adaptation to climate change.[9]

The movement of people from their homes, whether on a temporary or a permanent basis, will raise a range of challenges, particularly legal, normative and practical, on the handling of climate displacement, but it will also have resource and security implications.[10] At the opening of the 68th session of the UN General Assembly in September 2013, the President of the Republic of Nauru, who is currently chairing the Alliance of Small Island States (AOSIS), made a statement that attracted worldwide attention, calling for the appointment of a Special Representative on Climate and Security who would help expand the understanding of the security dimensions of climate change, and would become an invaluable asset in preventive diplomacy efforts and post-conflict situations.[11]

The magnitude of future displacement and its impact on the rights of those concerned, necessitate sustainable solutions, proactive and well-planned approaches, that are people-centred and rights-based.[12] As the United Nations Deputy High Commissioner for Human Rights has stated,

> Regrettably-and ... perilously, human rights have remained a peripheral concern in negotiations, discussions and research related to global warming. While the environmental, technical, economic, and more recently, developmental aspects of climate change have been explored, much less has been heard about its human rights dimension.[13]

The apprehension of the potential protection gap has to lead to a fundamental change of the existing modus operandi.[14] The Peninsula Principles are a significant step into this direction.

Climate displacement scenarios – legal implications and gaps

In order to be able to fully assess the impact and legal implications of climate change, it is crucial to identify potential displacement scenarios and to assess the character of the displacement as well as the needs for protection and assistance of those concerned.

The Representative of the Secretary-General on internally displaced persons (IDPs)[15] identified four main climate change disaster types,[16] which may coincide and/or overlap:

- hydro-meteorological disasters, which are projected to increase further in future and to lead to new and larger situations of displacement, and which often go hand in hand with destruction of property;
- general environmental degradation and slow-onset disasters, i.e. the deterioration of conditions of life and economic opportunities in affected areas, which will induce voluntary population movements in an initial phase, but could later change into forced, permanent displacement as areas become less hospitable due to desertification or rising sea levels;

- sinking small island States, which will prompt internal relocation and migration abroad, including forced and permanent displacement; and
- climate change-induced armed conflict and violence triggered by a decrease in vital resources, attributable to climate change.[17]

These categories can be related to a range of displacement scenarios. Depending on whether the resulting displacement is temporary or permanent in nature and on whether displacement occurs within the territory of a State or implies cross-border movements, with people remaining within the same region or moving to other continents, the following scenarios are identified:

- temporary displacement as it occurs after hurricanes, floods or storms
- permanent local displacement, if people are not able to return to their home because of irreversible changes in their living environment;
- permanent internal displacement;
- permanent regional displacement;
- permanent inter-continental displacement.[18]

In terms of protection, some of the disaster and displacement scenarios raise particular legal questions in connection with the applicable legal framework and States' responsibilities. Some movements prompted by climate change seem to fall within the classical refugee law framework: for example, refugee flows provoked by climate-induced armed conflict. International or regional refugee instruments and complementary forms of protection are applicable in this context, which also falls within UNHCR's protection mandate.[19]

Factual and legal difficulties arise in the event of forced cross-border population movements which do not result from internal violence or conflict. Individuals crossing international borders as a result of the erosion of their habitable areas do not fall within the existing refugee definition. Another lacuna exists in relation to populations who are forced to resettle as their State's territory ceases to exist. This will be the case for a number of sinking small island developing States (SIDS). Apart from the legal perspective, this also raises issues of preservation of cultural heritage and cultural identity. These aspects involving cross-border displacement will not, however, be dealt with here, since questions that touch upon the issues of statehood, statelessness and the protection of cross-border climate migrants can only be resolved in a multilateral framework and through a State-driven process aimed at identifying and defining existing gaps, a process which takes appropriate steps, where necessary, to adjust or further elaborate the legal framework on the basis of an international consensus. Meanwhile, as long as this gap has not been bridged, and until adequate responses have been developed to address these more general legal questions, attention should shift to what the new reality means in practice for the protection of human rights and how it can be addressed.

While climate-induced IDPs are in general better protected than cross-border refugees, there are practical, legal and normative gaps that still need to be addressed, notably as regards the definition of forced displacement, which may be invoked by such IDPs, as opposed to voluntary migration, and the question of return, since their displacement may be permanent in nature.[20] In order to address these challenges resulting from climate displacement in an adequate manner, there is a pressing need to develop appropriate national and international normative, institutional and implementation frameworks.

Climate change and human rights

In 1994, in its study on Human Rights and the Environment, the Sub-Commission on Prevention of Discrimination and Protection of Minorities[21] noted that:

> For many years environmental problems were almost exclusively considered from the standpoint of the pollution in one part of the world.... Acknowledgement of the link between the environment and human rights was fostered by an awareness of the global, complex, serious and multidimensional nature of environmental problems.[22]

Similarly, whereas the social and human dimensions of climate change as among the most important drivers of ecosystem changes have gained broad recognition over past years, the nexus between climate change and the effective enjoyment of human rights has only recently been acknowledged.[23]

The Inuit Petition,[24] filed with the Inter-American Commission on Human Rights in December 2005, alleging the responsibility of the United States of America for the effects of global warming which constitutes a violation of the rights of the Inuit, although rejected, is an important acknowledgement in this respect, as the Commission invited the Inuit Alliance in February 2007 to testify on the linkages.[25]

By the same token, the Malé Declaration,[26] adopted as a result of the SIDS Conference convened by the Maldives in November 2007, stresses that the environment provides the infrastructure for human civilization and that climate change poses the most immediate, fundamental and far-reaching global threat to the environment, to individuals and to communities. This constitutes the first explicit recognition of the impact of climate change on human rights at the international level. Moreover, the Malé Declaration calls for the cooperation of the Office of the UN High Commissioner for Human Rights (OHCHR) and the United Nations Human Rights Council in assessing the human rights implications of climate change.[27]

The Malé Declaration set in motion a process which culminated in the adoption by the Human Rights Council of resolution 7/23 on human rights and climate change in March 2008. With this resolution, the implications of climate change for the enjoyment of human rights were ascertained for the first time at the UN level, and the UN High Commissioner for Human Rights was commissioned to conduct a study on States' obligations under international human rights law to protect those rights from the effects of climate change.[28]

The report of the High Commissioner, which was presented a year later at the tenth session of the Human Rights Council in March 2009, concludes that climate change has a range of implications for the effective enjoyment of human rights, be it of a direct or indirect nature, as the impact of climate change on human rights may often translate into discriminatory policies and treatment. It also outlines the different legal guarantees for those displaced as a result of climate change, differentiating between different forms of displacement

> Persons affected by displacement within national borders are entitled to the full range of human rights guarantees by a given State, including protection against arbitrary or forced displacement and rights related to housing and property restitution for displaced persons. To the extent that movement has been forced, persons would also qualify for increased assistance and protection as a vulnerable group in accordance with the Guiding Principles on Internal Displacement.[29]

While the negative effects of climate change on the realization of human rights are not contested, the key question that remains is whether this impact would qualify as a human rights violation.[30] The report also identifies one of the key challenges in relation to litigation, i.e. the attribution of climate change-related harm to acts or omissions of States, while recognizing the need for redress and adequate legal protection against climate change-related or induced violations, and affirms the need to further study protection mechanisms for climate change-displaced persons.[31] In terms of protection, notwithstanding these questions, the human rights obligations of States provide the affected individuals with important guarantees.[32]

This interrelationship between human rights and climate change was again explicitly acknowledged by resolution 10/4 which the Human Rights Council adopted on 25 March 2009 as a result of consideration of the High Commissioner's report.[33] Further to this resolution, a panel discussion on the relationship between climate change and human rights was held at the 11th session of the Council, to improve understanding of the implications of climate change on the full enjoyment of human rights, and to discuss the implications on policy-making and further development of international law.

Other human rights bodies have also considered the matter. The Committee on the Elimination of Discrimination against Women at its 44th session in August 2009 adopted a statement on gender and climate change, noting that from an examination of State Parties' reports, it is apparent that climate change does not affect women and men in the same way and has a gender-differentiated impact.[34]

Measures and actions to address the impact of climate change on the full enjoyment of human rights at the local, national, regional and international levels were further discussed during the 2010 Social Forum pursuant to Human Rights Council resolution 13/17. The Forum concluded, in line with the Cancún Adaptation Framework, that the movement of people should be seen as a possible and legitimate adaptation strategy, among other options, and recommended in this regard that

> the situation of persons displaced as a result of climate change be addressed, and that national Governments in disaster-prone countries invest in disaster risk reduction planning, mechanisms and procedures, as well as in other adaptation measures in the face of the already present negative impacts of climate change, and that the members of the international community increase their efforts in responding to international disasters and in investing in preparedness, adaptation and mitigation through sustainable development options, which may imply the transfer of best available technologies.[35]

Like the Sub-Commission in 1994,[36] it also recommended the creation of a new Special Procedures mandate.[37]

In September 2011, the Human Rights Council adopted its third resolution (resolution 18/22) on the issue of human rights and climate change. It reiterated 'its concern that climate change poses an immediate and far-reaching threat to people and communities around the world and has adverse implications for the full enjoyment of human rights'.[38] In implementation of this resolution, the High Commissioner organized a seminar in February 2012 on the issue of addressing the adverse impacts of climate change on the full enjoyment of human rights, with a view to following up on the call for respecting human rights in all climate change-related actions and policies and forging stronger cooperation between the human rights and climate change communities. The UN High Commissioner for Human Rights emphasized that addressing the issue 'requires considering how to internalize the human consequences of climate change from a rights-based perspective, rather than limiting ourselves to quantitative dimensions alone'.[39]

Suggestions made previously for the creation of a new mandate of Special Rapporteur for climate change and human rights were reiterated at the seminar.[40] The Council followed a somewhat different approach with its resolution 19/10 of March 2012, when it decided instead to establish a mandate on human rights and the environment, which will study the human rights obligations relating to the enjoyment of a safe, clean, healthy and sustainable environment, and promote best practices relating to the use of human rights in environmental policy-making. The key issue in the climate change debate, concerning this phenomenon's human rights nexus, remains the same when it comes to environmental questions at large. In his first report to the Human Rights Council, the new Independent Expert outlined the linkage between substantive and procedural rights and duties in the area of human rights and the environment, such as access to information, public participation, and access to legal remedies, and, like the Representative of the Secretary-General on IDPs (see below),[41] emphasized the key role of compliance with procedural rights in terms of impact on the protection of substantive rights.[42]

Other Special Procedures of the Human Rights Council also significantly contributed to analyses of the linkages between human rights and climate change. In 2011, the Representative of the Secretary-General on IDPs, for instance, devoted his report to the General Assembly to climate change and human rights.[43] The Special Rapporteur on adequate housing as a component of the right to an adequate standard of living, and on the right to non-discrimination in this context, not only prepared a report on climate change[44] but also integrated the climate displacement perspective in a number of other thematic and mission reports.[45]

While these developments seem to demonstrate a growing international recognition of the effects of climate change on the enjoyment of human rights, a common and shared understanding on this issue has not, thus far, emerged. Some invoke practical constraints to a human rights-based approach to climate change such as path-dependency,[46] while others seem concerned that scientific debates will become 'politicized' through the human rights angle.[47]

A rights-based approach to climate displacement

Climate-related human rights concerns are manifold, as its consequences may impact, among others, the rights to life, health, education, access to safe drinking water and an adequate standard of living, including housing.[48] In this context, it is noteworthy that the need-centred approach to development has been largely replaced by a rights-based approach, which sets the frame for consistent and anticipatory rights-based climate displacement policy and action.[49] A human rights-based approach links the assessment of the impact of climate change on individuals and groups with the accountability framework based on the human rights obligations of States to respect, protect and fulfil human rights.

The Guiding Principles on Internal Displacement,[50] adopted in 1998, provide that 'at the minimum, regardless of the circumstances, and without discrimination, competent authorities shall provide internally displaced persons with and ensure safe access to … basic shelter and housing' (Principle 18), and also stipulate a post-displacement right to restitution of property (Principle 29). More specific to housing and property rights, similar Principles are detailed in the United Nations Principles on Housing and Property Restitution for Refugees and Displaced Persons, commonly referred to as the Pinheiro Principles,[51] endorsed in 2005 by the United Nations Sub-Commission for the promotion and protection of human rights. They encapsulate and restate existing legal norms and obligations on the right to restitution of housing, land and property (HLP) as well as the right to return, and they provide *all* refugees and displaced persons with a distinct and genuine individual right to claim back their homes, land or property.[52]

While the Pinheiro Principles provide for protection for *all* displaced persons irrespective of the causes of displacement and their location, i.e. including as a result of changing climatic conditions, they are not specifically designed to address situations of climate displacement and focus particularly on HLP rights. The Guiding Principles, by contrast, address the broader issue of internal displacement but do not, beyond the general Principle, provide practical guidance in terms of the handling of climate displacement situations.[53]

Another set of Principles, the so-called Nansen Principles, developed at the Nansen Conference on Climate Change and Displacement in June 2011, seek specifically to guide responses to the challenges raised by cross-border displacement in the context of climate change and underscore the need for a human rights-based approach (Principle I). Principle VII emphasizes that 'the existing norms of international law should be fully utilized' and at the same time recognizes that 'normative gaps [need to be] addressed'.[54]

The imperative of a rights-based rather than a needs-based approach was also recognized by the Representative of the Secretary-General on IDPs, who noted in his 2011 report to the General Assembly that it was essential that adaptation frameworks be comprehensive in nature, that they adopt a human rights-based approach, and that they be adequately supported.[55]

The recognition of the need to address climate displacement in a comprehensive rights-based manner finally led to the elaboration of the Peninsula Principles. These Principles explicitly build on the Guiding Principles and contextualize them, and also acknowledge, among others, the Pinheiro Principles and incorporate some of their provisions. By contrast to a number of proposals for a new legal instrument providing protection for people displaced by climate change,[56] the Peninsula Principles constitute a restatement of the law tailoring it to the specific needs of climate displacement, addressing existing gaps and formulating policy and action guidance.

In concrete terms, the Peninsula Principles seek to offer responses to massive displacement in the context of climate change. They grew out of a necessity to address climate displacement, and to restate and articulate human rights and other obligations deeply grounded in existing international public and human rights law. In an interplay with the Nansen Initiative,[57] which focuses on the needs of persons displaced across borders and aims at developing a protection agenda in this regard, the Peninsula Principles restate human rights law and principles focused on internal climate displacement and provide a comprehensive normative framework and minimum standards reflecting well-established good practice, within which the rights of climate-displaced persons can be addressed. These tenets not only provide and promote protection principles and legal obligations, but they also formulate concrete policy and management measures in responding to situations of climate displacement in a human rights-compatible manner throughout all stages, i.e. before, during and after a climate displacement situation.

The Principles arose from a climate displacement law project initiative, coordinated by Scott Leckie and David Hodgkinson working under the auspices of Displacement Solutions, aimed at enunciating and further developing the normative framework when it comes to the handling of climate displacement; they are the result of a year-long process of consultations.

The Peninsula Principles emphasize the common and correct view of climate-displaced persons as rights-holders. They outline the rights of individuals as well as communities who are facing or experiencing climate displacement, including those who lose their homes, land and livelihoods as a result of climate change. At the same time, the Principles set out the protection and assistance standards to be applied to climate-displaced persons and provide guidance, assistance and support to governments and local authorities on how best to address climate displacement in very practical and concrete terms, complying with their human rights and good governance obligations, so as to be able to pre-emptively manage and adequately react to climate displacement challenges. Most significantly, the Principles recognize that displacement situations invariably give rise to HLP-related issues,[58] and emphasize the importance of land in the resolution of climate displacement. They require States to identify and acquire suitable land as well as to develop relocation sites in order to provide viable and affordable land-based solutions to climate displacement. Emphasis is thereby placed on the habitability and feasibility of the identified alternatives.[59]

Conclusion

The human consequences of climate change include the erosion of the rights of those affected. Displacement may become a legitimate means to adapt to altered living conditions. States have a responsibility and obligation to address these challenges and need to take measures to alleviate the impact of climate change, particularly through the provision of adaptation assistance, including land allocation programmes and planned relocation as well as the provision of effective remedies in case of violation of HLP rights. The Peninsula Principles are specifically designed to respond to these challenges.

Notes

1 *Webster's New World Medical Dictionary*, 3rd Edition, Wiley, 2008, p. 19; Dorland's *Illustrated Medical Dictionary*, 32nd Edition, Elsevier, 2011, *inter alia* at p. 75.
2 The IPCC was established in 1988 by the World Meteorological Organization and the UN Environment Programme, to assess the information relevant to the scientific basis of the risk of human-induced climate change, its potential impact and possible response strategies.
3 O. Dun and F. Gemenne, Defining 'environmental migration', in: *Forced Migration Review*, Issue 31, October 2008, pp. 10–11.
4 IPCC, First Assessment Report, 1990.
5 International Organization for Migration, Migration and Climate Change, 2008, pp. 11ff., http://publications.iom.int/bookstore/free/MRS-31_EN.pdf; N. Myers, Environmental Refugees: An emergent security issue, 13th Economic Forum, Prague, May 2005; N. Stern (ed.), *The Economics of Climate Change: The Stern Review*, The Stationery Office, 2006, p. 3; see also K. Warner, Climate Change Induced Displacement: Adaptation Policy in the Context of the UNFCCC Climate Negotiations, 2011, p. 2: www.iom.int/cms/climateandmigration#; see also O. Brown, The numbers game, *Forced Migration Review*, Issue 31, October 2008, pp. 8–9; and K. Warner and F. Laczko, A global research agenda, *Forced Migration Review*, Issue 31, October 2008, pp. 59–61.
6 See E. Piguet, A. Pécoud and P. de Guchteneire, Migration and climate change, *Refugee Survey Quarterly*, Vol. 30, No. 3, 2011, pp. 12ff.
7 See IPCC, Fourth Assessment Report, 2007.
8 Ibid., p. 102; For an overview of various policy proposals, see S. C. McAnaney, Sinking islands? Formulating a realistic solution to climate change displacement, *New York University Journal of International Law and Politics*, October 2012, pp. 1173–1209; similarly, K. M. Wyman, Responses to climate migration, *Harvard Environmental Law Review*, Vol. 37, 2013, pp. 167–216; S. Martin, Climate change, migration, and governance, *Global Governance*, 16, 2010, 397–pp. 414.
9 UNFCCC, Outcome of the work of the Ad Hoc Working Group on Long-term Cooperative Action under the Convention, CP.16, 2010, para. 14(f), http://unfccc.int/files/meetings/cop_16/application/pdf/cop16_lca.pdf.
10 A. Morton, P. Boncour and F. Laczko, Human security policy challenges, *Forced Migration Review*, Issue 31, October 2008, pp. 5–7, at p. 5.
11 Statement by His Excellency the Honourable Baron Divavesi Waqa MP, President of the Republic of Nauru, at the General Debate of the 68th Session of the United Nations General Assembly, Thursday, 26 September 2013, United Nations, New York. On 20 July 2011, the Security Council also held a debate on the possible security implications of climate change, in which it was noted that climate change could aggravate or amplify existing security concerns and give rise to new ones, particularly in already fragile and vulnerable nations: see www.un.org/News/Press/docs//2011/sc10332.doc.htm. Also of note, the Secretary-General announced his intention during the General Debate to convene a third high-level event on climate change in 2014.
12 Morton *et al.*, Human security policy challenges, p. 6; S. Leckie, Human rights implications, *Forced Migration Review*, Issue 31, October 2008, pp. 18–19.
13 K. Kang, Climate Change, Migration and Human Rights: Address by Ms Kyung-wha Kang, Deputy High Commissioner for Human Rights Office of the United Nations High Commissioner for Human Rights, at the Conference on Climate Change and Migration: Addressing Vulnerabilities and Harnessing Opportunities, Geneva, February 2008.

14 See *inter alia* B. Docherty and G. Tyler, Confronting a rising tide: A proposal for a convention on climate change refugees, *Harvard Environmental Law Review*, Vol. 33, 2009, pp. 349–403; F. Biermann and I. Boas, Preparing for a warmer world: towards a global governance system to protect climate refugees, *Global Environmental Politics*, Vol. 10, No. 1, February 2010, pp. 60–88; D. Hodgkinson *et al.*, Towards a convention for persons displaced by climate change: key issues and preliminary responses, *The New Critic*, Issue 8, September 2008, www.ias.uwa.edu.au/new-critic/eight/?a=87815; A. Williams, Turning the tide: recognizing climate change refugees in international law, *Law & Policy*, Vol. 30, No. 4, October 2008, pp. 502–529; J. B. Cooper, Environmental refugees: meeting the requirements of the refugee definition, *N.Y.U. Environmental Law Journal*, Vol. 6, 1997–1998, pp. 480–529.

15 Now Special Rapporteur on Internally Displaced Persons: see Human Rights Council resolution 14/6 of 17 June 2010.

16 The issue of the protection of persons in the event of disasters is also on the programme of work of the International Law Commission which seeks to address the need for enhanced legal regulation: see *inter alia* A/CN.4/662.

17 The Representative of the Secretary-General (RSG) on the Human Rights of Internally Displaced Persons, W. Kälin, United Nations Inter-Agency Standing Committee Group on Climate Change, Background Paper, Displacement Caused by the Effects of Climate Change: Who will be affected and what are the gaps in the normative frameworks for their protection?, 10 October 2008, p. 2.

18 S. Leckie, Climate-Related Disasters and Displacement: Homes for Homes, Lands for Lands, in: J. M. Guzman, G. Martine, G. McGranahan, D. Schensul and C. Tacoli (eds), *Population Dynamics and Climate Change*, UNFPA and IIED, 2009, pp. 119–132, at Pp. 122ff.

19 UNHCR, Climate Change, Natural-Disasters and Human Displacement: A UNHCR Perspective, 14 August 2009, pp. 2ff.

20 K. Koser, Gaps in IDP protection, *Forced Migration Review*, Issue 31, October 2008, p. 17; R. Zetter, Legal and normative frameworks, *Forced Migration Review*, Issue 31, October 2008, pp. 62–63.

21 Renamed by the Economic and Social Council (ECOSOC) in 1999 in Sub-Commission on the Promotion and Protection of Human Rights; see also www2.ohchr.org/english/bodies/subcom/index.htm.

22 E/CN.4/Sub.2/1994/9, para. 6.

23 See J. Knox, Climate change and human rights law, *Virginia Journal of International Law*, Vol. 50, No. 1, pp. 164–218, at pp. 190ff.

24 Inuit Circumpolar Conference (ICC) petition seeking relief from violations resulting from global warming caused by acts and omissions of the United States, 7 December 2005.

25 Letter from the Assistant Executive Secretary of the Inter-American Commission on Human Rights of the Organization of American States to the Inuit Alliance, Earthjustice and the Centre for International Environmental Law of 1 February 2007, www.ciel.org/Publications/IACHR_Response_1Feb07.pdf.

26 AG/RES.2429 (XXXVIII-O/08), Human Rights and climate change in the Americas; Malé Declaration on the Human Dimension of Global Climate Change, 14 November 2007.

27 AG/RES.2429 (XXXVIII-O/08), Human Rights and climate change in the Americas; Malé Declaration on the Human Dimension of Global Climate Change, 14 November 2007.

28 A/HRC/RES/7/23.

29 A/HRC/10/61, para. 57.

30 A/HRC/10/61, para. 69; see also E. Cameron, Human rights and climate change: moving from an intrinsic to an instrumental approach, *GA Journal of International and Comparative Law*, Vol. 38, 2010, pp. 673–716, at pp. 701ff.

31 A/HRC/10/61; see also M. S. Chapman, Climate Change and the Regional Human Rights Systems, *Sustainable Development Law & Policy*, Vol. 10, No. 37, 2009–2010, pp. 37–39; A. Boyle, Human rights and the environment: where next?, *The European Journal of International Law*, Vol. 23, No. 3, pp. 613–642, at pp. 613 ff; Council of Europe, *Manual on Human Rights and the Environment*, 2nd Edition, Council of Europe, 2012; P. Stephens, Applying human rights norms to climate change: the elusive remedy, *Colorado Journal of International Environmental Law and Policy*, Vol. 21, 2010, pp. 49–83.

32 A/HRC/10/61, para. 71.

33 A/HRC/RES/10/4.

34 www.ohchr.org/Documents/HRBodies/CEDAW/Statements/StatementGender-ClimateChange.pdf.

35 A/HRC/16/62, para. 60 (f).

36 E/CN.4/Sub.2/1994/9.

37 A/HRC/16/62.

38 A/HRC/RES/18/22.

39 A/HRC/20/7, para. 64; see also DPI/2483, OHCHR, United Nations Joint Press Kit for the Bali Climate Change Conference, 3–14 December 2007: The Human Rights Impact of Climate Change, November 2007.

40 Such a mandate had already been recommended in 1994 by the former Sub-Commission on the Prevention of Discrimination and Protection of Minorities, see E/CN.4/Sub.2/1994/9; see also above the related discussions at the Social Forum.

41 A/66/285, para. 81 *et seq.*

42 A/HRC/22/43, para. 40 *et seq.*; see also Statement of the European Union Delegation at the 22nd session of the Human Rights Council delivered during the interactive dialogue with the Independent Expert on the issue of human rights obligations relating to the enjoyment of a safe, clean, healthy and sustainable environment, 7 March 2013; Boyle, Human rights and the environment, pp. 621ff.

43 A/66/285; A/HRC/13/21.

44 A/64/255.

45 See the Special Rapporteur's report on her mission to the Maldives in February 2009, A/HRC/10/7/Add.4 and A/HRC/13/20/Add.3; see also A/HRC/16/42 (Post Conflict and Post Disaster Reconstruction and the Right to Adequate Housing); A/66/270 (The Right to Adequate Housing in Disaster Relief Efforts).

46 National Office of Oceanic and Atmospheric Research, Global Warming and Hurricanes, www.oar.noaa.gov/spotlite/archive/spot_gfdl.html.

47 M. Limon, Human rights and climate change: constructing a case for political action, *Harvard Environmental Law Review*, Vol. 33, pp. 439–476, at pp. 459ff.

48 For a comprehensive analysis of the effects of climate change on human rights, see A/HRC/10/61.

49 W. Sachs, Climate change and human rights, *Development*, Vol. 51, No. 3, 2008, pp. 332–337.

50 E/CN.4/1998/53/Add.2.

51 E/CN.4/Sub.2/2005/17.

52 FAO/IDMC/OCHA/OHCHR/UN-Habitat/UNHCR, Handbook, Housing and Property Restitution for Refugees and Displaced Persons, Implementing the 'Pinheiro Principles', March 2007, pp. 16ff.; K. Hassine, Regularizing Property Rights in Kosovo and Elsewhere, WiKu.

53 E/CN.4/1998/53/Add.2.

54 www.regjeringen.no/upload/UD/Vedlegg/Hum/nansen_prinsipper.pdf.

55 A/66/285.

56 B. Docherty and G. Tyler, Confronting a rising tide: A proposal for a convention on climate change refugees, *Harvard Environmental Law Review*, Vol. 33, 2009, pp. 349–403; F. Biermann and I. Boas, Preparing for a warmer world: towards a global governance system to protect climate refugees, *Global Environmental Politics*, Vol. 10, No. 1, February 2010, pp. 60–88; D. Hodgkinson *et al.*, Towards a Convention for Persons Displaced by Climate Change; J. Bétaille *et al.*, Projet de Convention relative au statut international des déplacés environnementaux, *Revue Européenne de Droit de l'Environnement*, 2008, pp. 381–393.

57 See W. Kälin, From the Nansen Principles to the Nansen Initiative, *Forced Migration Review*, Issue 41, December 2012, pp. 48–49.

58 UN IASC, Protecting Persons Affected By Natural Disasters: IASC Operational Guidelines on Human Rights and Natural Disasters, 2006.

59 See Peninsula Principle 11; see also S. Leckie, *Finding Land Solutions for Climate Displacement: A Challenge Like Few Others*, Displacement Solutions, 2013.

3   A brief overview of the drafting of the Peninsula Principles

David Hodgkinson

From 16 to 18 August 2013 on the Mornington Peninsula in Victoria, Australia, an expert group comprised UN officials, judges, international lawyers, jurists, academics, NGO leaders, researchers, practitioners, consultants and activists met to consider an advanced draft of the Peninsula Principles on Climate Displacement within States – or what came to be called the Peninsula Principles. That draft was the result of more than two years of consultation on a global scale, during which time 22 substantive drafts of the Peninsula Principles were produced.

This chapter sets out the drafting assumptions, underpinnings and, most importantly, the history of the Principles. It sets out the ways in which the Principles were drafted, the methods employed, and the documents which served as the basis for many of the provisions.

Drafting assumptions and underpinnings

It was clear from the start that current protections under international law do not adequately provide for a number of the categories of persons likely to be displaced by climate change.[1] International refugee lawyers generally agree that those displaced by the effects of climate change would not be the subject of protection under the Refugee Convention.[2] The Refugee Convention, the most comprehensive articulation of refugee rights and State obligations, relies upon a restrictive definition of a refugee as someone with a 'well-founded fear of being persecuted for reasons of race, religion and nationality, membership of a particular social group or political opinion', and who is 'outside the country of his nationality'.[3] It would be difficult to establish that a person displaced by climate change has been 'persecuted', as required by the Refugee Convention and defined by the existing jurisprudence.

Another concern with seeking protection for such displaced persons under the Refugee Convention is that doing so would risk devaluing current protections for refugees.[4] Furthermore, to conflate the term 'refugee' such that it includes both those displaced by climate change and traditional refugees obscures fundamental differences of experience between the groups; most problematically, as a general rule the nexus between climate-displaced persons and their States will not have been severed through persecution. Such protections would also be insufficient; most displacement due to climate change will be internal, meaning most of those displaced will not cross national borders.

Proposals for some kind of legal instrument designed to address the problem of climate change displacement seek, in various ways and for various reasons, to link that instrument with the United Nations Framework Convention on Climate Change (UNFCCC).[5] Yet neither the UNFCCC process nor its Kyoto Protocol[6] contemplates or addresses the issue of displacement. The UNFCCC has limitations as a framework for dealing with climate change displacement. Displacement is not its focus; its concerns lie instead in the promotion of climate mitigation activities. Its structure and institutions are not designed to address displacement and the issues associated with it. Moreover, the UNFCCC cannot easily be altered in order to accommodate persons displaced by climate change; dealing with existing provisions is already problematic.

These arguments are put succinctly by Docherty and Giannini.[7] While they observe that the UNFCCC does apply directly to climate change, they also note that it has legal limitations for dealing with climate change displacement. As an environmental law treaty, the UNFCCC primarily concerns State-to-State relations; it does not discuss duties that States have to individuals or communities, such as those laid out in human rights or refugee law. It is also preventive in nature and less focused on the remedial actions that are needed in a refugee context.

Like the refugee regime, the UNFCCC was not designed for, and to date has not adequately dealt with, the problem of displacement generated by the effects of climate change. Further, to the extent that the UNFCCC does consider displacement effects from climate change, its focus is heavily on international movement. A UNFCCC 'non-paper' is one used by State parties as a starting point for work on negotiating texts, as well as a vehicle for comments and proposing revisions. Non-paper 41 on climate change adaptation, produced after the seventh session of the Ad Hoc Working Group on Long-Term Cooperative Action (AWG-LCA) under the UNFCCC in Barcelona in October 2009, refers to migration and displacement as follows:

> All Parties [shall] [should] jointly undertake action under the Convention to enhance adaptation at the international level, including through … (b) Activities related to migration and displacement or planned relocation of persons affected by climate change, while acknowledging the need to identify modalities of interstate cooperation to respond to the needs of affected populations who either cross an international frontier as a result of, or find themselves abroad and are unable to return owing to, the effects of climate change.[8]

While the Norwegian Refugee Council argues that '[i]t is important that the reference [in the non-paper] to the international level is not interpreted to mean that only cross-border movements are addressed',[9] it is clear that, for the AWG-LCA, the emphasis is on international displacement. Again, then, most displacement will likely occur within national borders. Internal rather than cross-border displacement is therefore the focus of the Peninsula Principles.

Existing proposals to address the displacement problem

Proposals for a new instrument providing for people displaced by climate change have been advanced in varying levels of detail by Docherty and Giannini,[10] Biermann and Boas,[11] Williams,[12] Bétaille *et al.*,[13] Hodgkinson *et al.*[14] and others.[15] All of the proposals agree that those displaced by climate change do not fall within the scope of the existing refugee regime created by the Refugee Convention. However, they differ as to the most appropriate instrument to tackle that problem, and the scope and detail of that instrument. The drafters of the Peninsula Principles acknowledged at various points in the drafting the contributions made by these authors, and considered and incorporated aspects of their proposals.

The drafting history in brief

The first draft of the Peninsula Principles was produced on 5 May 2013; 21 subsequent drafts followed ahead of the final drafting meetings held in August 2013. Each of the 22 drafts produced during the drafting process was discussed in telephone conversations and teleconferences, by skype and email. Unsurprisingly, each draft grew in length as (a) original provisions were either deleted or fleshed out and additional provisions added; (b) more primary and secondary source documents were consulted; and (c) academics and practitioners provided draft clauses and further ideas.

Those providing comments include legal practitioners, academics (from law and a number of other disciplines), activists, judicial officers, non-governmental organisations, non-profit associations and politicians (in their private capacity). Government and ex-government officials (also in their private capacity) also provided comments and suggested amendments to various drafts. An online video message call for comments was made by Displacement Solutions, inviting anyone over a six-week period to submit their views on the document as it looked prior to the drafting session. In total, hundreds of comments were provided from a diverse array of people, all of which were considered in the preparation of the Meeting Draft that was presented to the final drafting meeting in August 2013.

In drafting and then amending the Peninsula Principles, a number of instruments were consulted and formed the basis for a number of the Principles. Rights-based instruments consulted included the Universal

Declaration of Human Rights, the International Covenant on Economic, Social and Cultural Rights, the International Covenant on Civil and Political Rights and the Vienna Declaration and Programme of Action. Climate change instruments consulted were the UNFCCC and its Kyoto Protocol. In terms of refugees, the 1951 Convention relating to the Status of Refugees (including its 1967 Protocol), the 1969 Organisation of African Unity Convention Governing the Specific Aspects of Refugee Problems in Africa, the 1984 Cartagena Declaration on Refugees and the UN Principles on Housing and Property Restitution for Refugees and Displaced Persons were all reviewed. On internal displacement, the United Nations Guiding Principles on Internal Displacement and the African Union Convention for the Protection and Assistance of Internally Displaced Persons in Africa (the Kampala Convention) proved especially useful. Indeed, in the course of drafting the Peninsula Principles it became clear, at least to this author, that arguments in favour of using such *existing* instruments to address the climate displacement problem have some merit (see e.g. the United Nations Guiding Principles on Internal Displacement). On cross-border displacement, the Nansen Initiative was consulted. Other miscellaneous documents drawn upon included the Inter Agency Standing Committee (IASC) Operational Guidelines on the Protection of Persons in Situations of Natural Disasters and the Hyogo Framework for Action.

Final amendments to the Meeting Draft

At the Mornington Peninsula meeting, an extraordinarily rich and intense discussion, analysis and debate took place on the entire Meeting Draft. Every word, every line, every paragraph was scrutinised by the expert group and a series of fascinating exchanges took place that led ultimately to the final text of the Peninsula Principles. Many of these discussions are examined in the chapters that follow. One particularly extensive discussion focused on how to define the term 'climate displaced persons' which was amended during the final meetings from 'individuals, households or communities in urban and/or rural areas who face a real risk of or experience climate displacement' to such persons 'who face or experience climate displacement'. Principle 3 was thus, for instance, expanded to include the following provisions:

> States shall not discriminate against climate displaced persons on the basis of their potential or actual displacement, and should take steps to repeal unjust or arbitrary laws and laws that otherwise discriminate against, or have a discriminatory effect on, climate displaced persons.
> ...
> (c) States should ensure that climate displaced persons are entitled to and supported in claiming and exercising their rights and are provided with effective remedies as well as unimpeded access to the justice system.

Similarly, a decision was affirmed that the Peninsula Principles should only address internal displacement and that cross-border displacement matters were best left to the Nansen Initiative and various academics who were focusing on that issue. The rationale for focusing on internal displacement is captured in the recitals of the text:

> NOTING that while climate displacement can involve both internal and cross-border displacement, most climate displacement will likely occur within State borders; ...

> NOTING the work of the Nansen Initiative on disaster-induced cross-border displacement and its 'protection agenda' for people displaced across borders due to natural disasters and the adverse effects of climate change; [and]

> NOTING that these Peninsula Principles, being confined to climate displacement within States, necessarily complement other efforts to address cross-border displacement, such as the Nansen Initiative ...

The Guiding Principles on Climate Change Displacement (draft 1) had become, by the final draft negotiating stage, the Peninsula Principles on Climate Displacement within States. Deletion of 'Change' was made in part because of causation issues. McAdam notes[16] that it is 'conceptually problematic and empirically flawed in most cases to suggest that climate change alone causes migration' and that 'it would seem both practically impossible and conceptually arbitrary to attempt to differentiate between those displaced people who deserve "protection" on account of climate change and those who are victims of "mere" economic or environmental hardship'.[17]

Causation is clearly a difficult issue, but similarly complex questions of causation have been resolved in other contexts; complexity of a pursuit does not necessarily render that pursuit any the less worthwhile.

The issue of arbitrariness raised by McAdam appears to have been resolved by the UNFCCC in its focus on adaptation assistance or 'protection' to address the effects of climate change; State parties agree specifically to cooperate in preparing for adaptation to the impacts of climate change. More so than other worldwide global economic and environmental problems, there is some international recognition that responsibility for the effects of human-induced climate change is global and, in particular, that it rests with developed countries. It may be that the international community is more likely to provide assistance to communities displaced by climate change rather than by other economic and environmental factors. While this may be arbitrary in some cases, it can be argued that it represents an opportunity to provide assistance to affected populations that should not be ignored. Nonetheless, McAdam (who kindly provided useful comments on early drafts of the Principles) also notes that '[S]tates presently seem to lack the political will to negotiate a new instrument requiring them to provide international protection to additional groups of people' and asks why States would 'be willing to commit to, and realize protection for, people displaced by climate change'.[18]

States may indeed lack political will to do this, but they lack political will in an almost infinitely large range of other important areas as well. It is very much hoped that the Peninsula Principles will provide an impetus to States to understand the seriousness of the climate displacement issue, and then act to prevent and resolve it in the best manner possible.

Notes

1 Bonnie Docherty and Tyler Giannini, Confronting a Rising Tide: A Proposal for a Convention on Climate Change Refugees, 33 HARV. ENVTL. L. REV. 349 (2009) at 358; Office of the High Commissioner for Human Rights, *Report of the Office of the United Nations High Commissioner for Human Rights on the Relationship Between Climate Change and Human Rights*, p. 19, UN Doc. A/HRC/10/61 (15 January 2009); Walter Kälin, *Displacement Caused by the Effects of Climate Change: Who Will Be Affected and What Are the Gaps in the Normative Framework for Their Protection?* (2008), available at www.brookings.edu/papers/2008/1016_climate_change_kalin.aspx. For a more comprehensive consideration of the current international law regimes that may be applicable to climate change displacement, see David Hodgkinson *et al.*, 'The Hour When the Ship Comes In': A Convention for Persons Displaced by Climate Change' 36(1) MONASH UNIVERSITY L. REV. 69 (2010), and Angela Williams, Turning the Tide: Recognizing Climate Change Refugees in International Law 30 LAW & POL'Y 502 (2008). For a discussion of the appropriateness of the United Nations Security Council as a forum for addressing climate change displacement, see Frank Biermann and Ingrid Boas, *Preparing for a Warmer World: Towards a Global Governance System to Protect Climate Refugees* 21 (Global Governance, Working Paper No. 33, 2007).
2 See Williams, Turning the Tide, pp. 507–508.
3 Convention Relating to the Status of Refugees art 1A(2), 22 April 1954, 189 UNTS 150.
4 David Keane, The Environmental Causes and Consequences of Migration: A Search for the Meaning of Environmental Refugees 16 GEO. INT'L ENVTL. L. REV. 209, 214–217 (2004).
5 United Nations Framework Convention on Climate Change, 21 March 1994, 1771 UNTS 107.
6 Kyoto Protocol to the United Nations Framework Convention on Climate Change, 16 February 2005, 2303 UNTS 148.
7 Docherty and Giannini, Confronting a Rising Tide.
8 Ad Hoc Working Group on Long-Term Cooperative Action under the Convention, *Contact Group on Enhanced Action on Adaptation and its Means of Implementation* 3–4 (Non-paper No. 41, Draft Text, 2009), available at www.stakeholderforum.org/fileadmin/files/Non-Paper%2041.pdf. See also Norwegian Refugee Council, Climate Changed: People Displaced 23 (2009).
9 Norwegian Refugee Council.
10 Docherty and Giannini, Confronting a Rising Tide.
11 Biermann and Boas, *Preparing for a Warmer World*.

12 Williams, Turning the Tide.
13 Julien Bétaille *et al.*, Draft Convention on the International Status of Environmentally-Displaced Persons, 4 REVUE EUROPEENNE DE DROIT DE L'ENVIRONNEMENT 395 (2008).
14 Hodgkinson *et al.*, 'The Hour When the Ship Comes in'.
15 See ENVTL JUSTICE FOUND., No Place Like Home: Where Next for Climate Refugees? (2008); Md Shamsuddoha and Rezaul Karim Chowdhury, Climate Change Migrants, 74 TIEMPO 1, 3 (2010).
16 Jane McAdam, Swimming Against the Tide: Why a Climate Change Displacement Treaty is Not the Answer, 23 INT'L J. REFUGEE LAW 2, 4 (2011).
17 Ibid.
18 Ibid.

4 The Preamble

Bruce Burson

Introduction

In August 2013, a group of concerned human rights lawyers, displacement experts and civil society activists met at Mornington Peninsula in Victoria, Australia, to discuss the issue of climate change-related displacement within states. They were welcomed onto the land by a representative of the aboriginal people of the area. The group were informed by the representative that aboriginal oral traditions referred to a vast and open plain, over which their ancestors roamed. That plain now lies beneath Phillip Bay, a vast and open stretch of water, and the people walk elsewhere: a vivid and powerful reminder that change in the physical environment has always caused people to move for security, to protect livelihoods and to preserve their way of life.

The coming together of this group to draft the Peninsula Principles rested on two shared beliefs. First, that due to anthropogenic influences, changes in the planetary climate system will mean that future generations will be living in a warmer world that will impact increasingly negatively upon their lives and livelihoods. Second, that the negative impacts of climate change will increasingly reduce the ability of affected individuals, households and communities to remain living in their homes, and will hence contribute to population movements.

The publication of the Peninsula Principles in August 2013 was bookended by two major reports of the Intergovernmental Panel on Climate Change (IPCC). These dealt, respectively, with the state of scientific knowledge surrounding observed changes in the climate and the effect of those changes on human and eco-systems.

The science

The scientific backdrop was painted in the second of the IPCC reports, publication of which followed shortly after the publication of the Peninsula Principles. Titled *Climate Change 2013: The Physical Science Basis*, the report was issued by IPCC Working Group I as its contribution to the Fifth

Assessment Report of the IPCC. Its 'Summary for Policymakers' noted that warming of the climate system is unequivocal, and many of the observed changes since the 1950s are unprecedented over decades to millennia (IPCC, 2013: 4). As to the driver of recent changes in the climate system, Working Group I are in little doubt, noting that 'it is extremely likely that human influence has been the dominant cause of the observed warming since the mid-20th century' (IPCC, 2013: 17).

Summarising the data, the report concludes, *inter alia*, that atmospheric concentrations of carbon dioxide (CO_2), and other greenhouse gases have increased to levels unprecedented in at least the last 800,000 years. There has been a 40 per cent increase in concentration of CO_2 since pre-industrial times, primarily from fossil fuel emissions, but also from net land use changes. The ocean has absorbed about 30 per cent of the emitted carbon dioxide, causing ocean acidification. Over the period 1901–2010, global mean sea-level rose by 0.19 m. Further, there was high confidence that the rate of sea-level rise since the mid-nineteenth century had been larger than the mean rate during the previous two millennia. It is very likely that sea-level rise in the twenty-first century will exceed that observed previously due to the combined effects of ocean warming and glacial and ice-shelf melt. It is virtually certain that, from 1971 to 2010, the upper ocean warmed to a depth of 700 m, and very likely that Arctic sea-ice will continue to shrink and thin.

A recent study has found that, unless greenhouse gas emissions are curbed, there is potential for planetary warming in the magnitude of 4°C by 2100 (Sherwood *et al.*, 2014). This would not simply be dangerous, but 'catastrophic' (Carrington, 2013). A recent report commissioned by the World Bank, *Turn Down the Heat: Why a 4°C Warmer World Must Be Avoided* considered the likelihood of global mean rise in surface temperature by 4°C. It notes that even if emission pledges made in Copenhagen and Cancún were implemented and met, this would still 'place the word on a trajectory for global mean warming of well over 3°C', with a 20 per cent chance of exceeding 4°C by 2100 (World Bank, 2012: 23). If even these modest pledges are not met, the chance of a 4°C warming by 2100 increases to over 40 per cent and, assuming a business-as-usual emissions pathway, creates a 10 per cent chance that this threshold will be reached as early as 2070.

The report examined the implications of a world warmed by 4°C, and identified a number of extremely severe risks for both human and eco-systems (World Bank, 2012: xi–xv). These include a likely regional extinction of entire coral reef systems, removing a natural source of protection against storm surges and erosion, and having a profound impact on the local eco-system and the people who depend on it for food, income and for tourism purposes. Global mean sea-level would rise by 0.5–1 m. However, given that the increase will be higher in tropical latitudes, there will be asymmetrical impacts with the impact being disproportionately felt in developing countries, where large populations live in coastally situated cities. Finally, the risk of extremes of both rainfall and drought are expected to double in magnitude when compared to a 2°C warmer world.

The human impacts of these observed phenomena were considered in the earlier of the two UNFCCC reports. In 2012, the special report *Managing the Risks of Extreme Events and Disasters to Advance Climate Change Adaptation* (the SREX report) was published. It outlines studies on a range of extreme weather and climatic phenomena that can be expected to increase the likelihood of population movements (see IPCC: 2012).

Cyclones/hurricanes/typhoons

It is estimated that 1.15 billion people live in cyclone-prone areas and that the number of persons physically exposed to cyclones is estimated to have increased from 73 million in 1970 to approximately 123 million in 2010 (Peduzzi *et al.*, 2011). Taking into account projected global population growth to 2030, even assuming a constant hazard (i.e. without factoring in the influence of climate change), it is expected that the average number of persons exposed annually to tropical cyclones will increase by 11.6 per cent over this period, with Asia having over 90 per cent of the total exposed population (Handmer *et al.*, 2012: 240–241).

Flooding

Approximately 800 million people currently live in flood-prone areas. Of these approximately 70 million are exposed to floods annually (UNISDR, 2011). The SREX report observes that, while there is considerable uncertainty in the magnitude, frequency and direction of changes to riverine flooding, nevertheless, even assuming a constant hazard, it is expected that population growth will increase exposure to flooding (Handmer *et al.*, 2012).

It has been recognised that climate change impacts could be a particularly important driver of migration in Asia, home to a sizeable portion of the global population (Hugo and Bardsley, 2014: 22). Depending on the scenario, the population living in urban floodplains in Asia may rise from 30 million in 2000 to between 83 and 91 million by 2030, and between 119 and 188 million in 2060 (Foresight, 2011: 110). Substantial parts of metropolitan Mumbai, a city with over 20 million inhabitants, are situated below sea-level and are already prone to flooding. As regards China, it is expected that, by 2030, more than one billion of its citizens will live in cities, and that 221 cities will have more than one million inhabitants. More than one-third of the Chinese population currently live in cities, including several megacities in coastal locations and riverine valleys susceptible to flooding and inundation (Roberts, 2010).

Thailand is located in one of the world's most disaster-prone regions and faces annual heavy rains and flooding.[1] Bangkok is particularly prone to flooding. The combined effects of sea-level rise, riverine flooding and storm surges could not only displace its inhabitants from its low-lying areas, but also affect the role of the city as a domestic migration destination and as an international migration hub in the context of regional labour movements (Asian Development Bank, 2013: 31).

Drought

The SREX report notes that there has been an overall increase in observed dryness in Africa with drought particularly affecting the Sahel region, the Horn of Africa and Southern Africa (Handmer *et al.*, 2012: 253). Changes in hydrology and water resources have led to increased water stress in both human and ecological systems. Some 25 per cent of the African population live in drought-sensitive conditions. Drought also has significant impacts in Asia where it causes yield loss in approximately 15 per cent of the region's rice-growing areas (Handmer *et al.*, 2012: 255).

Heat-waves

The SREX report notes an increase in the frequency and duration of heat-waves (Handmer *et al.*, 2012: 251–252). It notes that, during the 2003 heat-wave in Europe, tens of thousands of additional heat-related deaths were recorded, with those most affected being vulnerable sections of the population such as the elderly, the sick and the socially isolated. Heat extremes also claim lives in tropical countries. Studies have noted higher mortality rates on very hot days in large cities in low- and middle-income countries such as Bangkok, Delhi and Salvador (Brazil).

While different regions experience these phenomena in different combinations and to varying degrees, these various hazards are expected to increasingly affect the lives of many millions of people in coming decades.

Climate change as a mobility driver

The First Assessment Report of the IPCC in 1990 that maintained 'the greatest single impact of climate change could be on human migration' (IPCC, 1992). At that time, there were many uncertainties, as to both the causes of climate change and its potential impacts on the physical and human environment.

Spurred on by concern about the wider impacts of anthropogenically driven changes in the climate system, increasing attention has been focused by researchers and academics from a range of disciplines on the relationship between changes in the physical environment and population movement. In the 25 years since the statement in the IPCC First Assessment

Report was made, the number of studies and reports on migration and the environment has increased from three in 1990 to 134 in 2011, with more being published in the three years from 2008 to 2011 than in the entire period 1990–2007 (Piguet and Laczko, 2014: 5). As the body of empirical research has grown, uncertainties have narrowed. It has become increasingly clear from the evidence that there is a correlation between changes in the environment and human mobility generally, but that the relationship is not mono-causal in nature and is mediated through a range of other factors. Factors such as age, gender, socio-economic status, perception of risk and culture can also play significant roles in determining who moves where and when in response to environmental stressors. The comprehensive report into the issue of climate change and migration undertaken by the United Kingdom's Foresight project (Foresight, 2011: 9) summarises the current consensus as follows:

> Environmental change will affect migration now and in the future, specifically through its influence on a range of economic, social and political drivers which themselves affect migration. However, the range and complexity of the interactions between these drivers means that it will rarely be possible to distinguish individuals for whom environmental factors will be the sole driver.

While climate will seldom be the only factor driving population movement, studies indicate that environmental stressors can nevertheless be significant in driving population movement without necessarily being the sole or even dominant factor. As changes in the physical environment caused or exacerbated by climate change become more pronounced, it can reasonably be expected that environmental factors will assume a more prominent role in driving migration and displacement. The SREX report acknowledges the potential for a stronger positive nexus between population movement and environmental degradation in a warmer world:

> If disasters occur more frequently and/or with greater magnitude, some local areas will become increasingly marginal as places to live or in which to maintain livelihoods. In such cases, migration and displacement could become permanent and could introduce new pressures in areas of relocation. For locations such as atolls, in some cases it is possible that many residents will have to relocate.
>
> (IPCC, 2012: 16)

This potential is also acknowledged in the recently published report by IPCC Working Group II, *Climate Change 2014: Impacts, Adaptation, and Vulnerability*, which records a 'high agreement' that climate change is projected to increase the displacement of people over the twenty-first century (IPCC, 2014: 20).

Reflecting both the residual uncertainties in the climate science and the multi-causal nature of migration generally, estimates of numbers of persons who are expected to be forced to move due to the negative impacts of climate change vary wildly. What is very clear, however, is that natural disasters currently cause substantial numbers of fatalities and contribute to homelessness on a widespread scale. For example, flooding in Bangladesh in 1998 killed approximately 1,100 people and left 30 million homeless (Handmer *et al.*, 2012: 253–254). In 2013, Typhoon Haiyan in the Philippines is estimated to have damaged or destroyed one million homes, ruined the livelihoods of five million people, and affected in excess of 16 million people (International Federation of the Red Cross, 2014).

Disaster risk is a function of exposure and vulnerability to weather and climate events. Disasters happen when human or ecological systems are overwhelmed and unable to cope. After a review of the literature, the SREX report notes 'with high confidence' that settlement patterns, urbanisation and changes in socio-economic conditions have all influenced observed trends in exposure and vulnerability to climate extremes (Handmer *et al.*, 2012: 234). The settlement patterns of an increasing global population both increase and concentrate exposure to natural hazards and increase vulnerability.

There are a number of relevant aspects in the context of displacement within states.[2] First, more people are living in places of high exposure. In particular, there has been a significant increase in the number of persons living in coastal locations of low elevation at risk of inundation through sea-level rise or from storm surges, and also exposed to further harm, loss and damage through increased cyclonic wind speeds. For example, in Bangladesh, one of the world's most disaster-prone countries, between 1990 and 2000 the coastal population grew at approximately twice the national rate (McGranahan *et al.*, 2007). Overall, the global coastal population is predicted to grow from 1.2 billion in 1990 to an estimated 5.2 billion by 2080 (Nicholls *et al.*, cited in Oliver-Smith, 2011: 168).

A second feature of settlement patterns is urbanisation and, in particular, the trend towards living in large cities. This increases exposure to the impacts of extreme temperature events and combined with the heat island effect can pose risk to life for the elderly, the unwell and other vulnerable sections of the population (Maloney and Forbes, 2011).

A final relevant feature to note is that approximately one billion people live in informal settlements, which are growing at a faster rate than formal settlements (UN-HABITAT, 2008; UNISDR, 2011). Not only are informal settlements often located in places with relatively high exposure to natural hazards, but their inhabitants have increased vulnerability due to the poor quality of housing construction materials, limited assets and financial resources, and often limited state support and legal protection (Dodman and Satterthwaite, 2008).

The structure of the Preamble to the Peninsula Principles

The 18 Principles contained within the Peninsula Principles on Climate Displacement within States are preceded by a Preamble containing 24 Recitals. Like the Preamble to a treaty, the Recitals preceding the Peninsula Principles are important in that they give hue and colour to the body of the document to which they relate and, as such, can provide an aid to interpreting its contents.[3] The Preamble to the Peninsula Principles records the concerns of the drafters, identifies their sources of inspiration, and provides the context for the substantive provisions.

Of profound concern is that, despite the existence of 'technically and economically feasible' (World Bank, 2012: xiv) emission pathways which could limit warming to 2°C or below, regrettably it remains to be seen whether they are politically feasible. Furthermore, even if meaningful mitigation steps were to be taken now which limited warming to 2°C or below, some degree of climate change is all but inevitable given the lags in the planetary climate system. As IPCC Working Group I observes:

> Cumulative emissions of CO_2 largely determine global mean surface warming by the late 21st century and beyond. Most aspects of climate change will persist for many centuries even if emissions of CO_2 are stopped. This represents a substantial multi-century climate change commitment created by past, present and future emissions of CO_2.
>
> (IPCC, 2013: 19)

It is clear, then, that we will be living in an increasingly negative climate-affected world. It is simply a question of how and where the effects will be felt.

Commenting on the risks posed by a 4°C world, *Turn Down the Heat* provides a stark warning:

> With pressures increasing as warming progresses toward 4°C and combining with non climate-related social, economic, and population stresses, the risk of crossing critical social system thresholds will grow. At such thresholds existing institutions that would have supported adaptation actions would likely become much less effective or even collapse. One example is a risk that sea-level rise in atoll countries exceeds the capabilities of controlled, adaptive migration, resulting in the need for complete abandonment of an island or region. Similarly, stresses on human health, such as heat waves, malnutrition, and decreasing quality of drinking water due to seawater intrusion, have the potential to overburden health-care systems to a point where adaptation is no longer possible, and dislocation is forced.
>
> Thus, given that uncertainty remains about the full nature and scale of impacts, there is also no certainty that adaptation to a 4°C world is possible. A 4°C world is likely to be one in which communities, cities

and countries would experience severe disruptions, damage, and dislocation, with many of these risks spread unequally. It is likely that the poor will suffer most and the global community could become more fractured, and unequal than today. The projected 4°C warming simply must not be allowed to occur – the heat must be turned down. Only early, cooperative, international actions can make that happen.

(World Bank, 2012: xviii)

The Preamble to the Peninsula Principles reflect the drafters' concern that, given that we are already locked into a warmer world for the long-term future, and given the uncertainty surrounding the development of a meaningful global mitigation architecture which limits the warming to below what is considered to be a safe threshold, it is prudent to prepare for the implications of a warmer world in which increasing numbers of individuals, households and communities may be forced to leave their homes and land.

The essence of this concern is captured at Recital 19, as follows:

Realising the need for a globally applicable normative framework to provide a coherent and principled approach for the collaborative provision of pre-emptive assistance to those who may be displaced by the effects of climate change, as well as effective remedial assistance to those who have been so displaced, and legal protections for both.

The remainder of the Recitals and the body of substantive provisions contained within the Principles amplify this concern. Read as a whole, the Recitals reveal three overarching themes which underpin the drafters' desire to create the globally applicable normative framework referenced in Recital 19. These themes are:

1   a focus on internal displacement in the specific context of climate change;
2   protecting rights, and framing duties and responsibilities; and
3   where population movement is unavoidable, planned relocation is preferable to displacement.

Theme 1: A focus on internal displacement in the context of climate change

Human mobility is generally understood to comprise forms of movement characterised by certain temporal and spatial minimums. Mobility exists along a number of intersecting axes: temporary–permanent; forced–voluntary; and internal (domestic)–external (international). It is widely recognised that along the forced–voluntary axis, forms of human mobility exist in a spectrum of movement, with a substantial zone of transition in the middle. Voluntary migration is characterised by the presence of available choice between options and the possibility of return to the place of origin (Kälin, 2013). Displacement, on the other hand, as a form of forced migration, is characterised by compulsion and the absence of choice.

Forms of human mobility arising from natural disasters and the adverse impacts of climate change range from evacuation and short-term displacement through to permanent relocation of entire communities and, at certain thresholds of climate change, potentially permanent relocation of even entire national populations. Various typologies of movement linked to changes in the environment caused or exacerbated by climate have been put forward. Leckie (2013: 22), for example, looking at the 'forced' end of the mobility spectrum, proposes the following:

- *Temporary displacement or evacuation* – persons forced to leave their home for short period of time because of an event, but who are able to return once the event has ceased;
- *Permanent local displacement* – persons forced to leave their homes due to irreversible changes in the physical environment where they habitually reside but who move on a permanent basis to a site situated near to their former home or place of habitual residence (e.g. uphill).
- *Permanent internal displacement* – persons displaced inside a national border but to places situated some distance away from their place of former habitual residence.
- *Permanent regional displacement* – persons displaced across national borders to other states within the region because alternative land suitable for habitation within the state of origin is unavailable.
- *Permanent inter-continental displacement* – persons displaced outside national and regional locations who receive protection or are otherwise able to access livelihoods in states in another continent (e.g. a former colonial power).

A similar typology has been put forward by Campbell (2010), which distinguishes between 'proximate' and 'non-proximate' movement. Proximate movement comprises movement of a localised nature, either to another site on communally owned land or on others' communal land in the locality. Non-proximate movement comprises movement of both an internal and international nature, with the internal component describing movement between regions existing inside national boundaries.

Population movement resulting from the adverse impact of climate change will, to a significant extent, follow existing patterns and pathways, particularly in social settings characterised by a well-established culture of mobility. While at certain thresholds of climate change cross-border population movement may be the dominant form of movement in some geographical areas, in the short to medium term it is anticipated that the current trend of internal displacement resulting from natural disasters being will continue. Accordingly, Recital 6 records the clear focus of the Peninsula Principles on Climate Displacement within States. There are a number of factors contributing to this current pattern of largely internal displacement.

First, the international border may be physically difficult to cross. In the Pacific, for example, international borders are typically 'lines on a map drawn over a vast oceanic space' across which movement is difficult to undertake (Burson and Bedford, 2013). Even where borders may be relatively easy to reach, the increasing securitisation of the border and emphasis on interdiction of irregular cross-border movement – particularly prevalent policies among countries of the industrialised 'global north' – also constitute an impediment.

Second, the financial cost of moving what may be large distances can be prohibitively high. This can be a particularly acute impediment for members of minority communities living on marginal land who often face greater disaster risk. It has been observed that groups such as pastoralists, farmers and fishermen, who are the most dependent on natural resources and weather for their livelihoods and thus more exposed to environmental stressors linked to climate change, are often those with the least capacity to use migration as a means of adapting to these stressors. Some affected communities may lack the capacity to move at all, or to move anywhere but a short distance, leading to the possibility of 'trapped' populations (Foresight, 2011).

Disaster impacts may also reduce mobility in some instances by increasing the labour needs in the place of origin (Foresight, 2011). A number of studies reinforce this point. A study of environmentally induced migration in Nepal has indicated that the livelihood impacts of the environmental change had, in some instances, decreased the likelihood of cross-border population movement, as the increase in time required for maintaining livelihoods *in situ* in more challenging environmental circumstances decreased the amount of time available to plan and prepare for international population movement (Bohra-Mishra and Massey, 2011: 93). A similar argument has been postulated in relation to Sahelian migration where reductions in farm income owing to sustained periods of drought arguably reduced the potential for international migration by making it more difficult for affected households to raise the necessary costs (Leighton, 2011). Finally a study on the nexus between migration and drought in the Horn of Africa revealed that only those with a sufficient asset base or transferable skills made conscious decisions to move permanently or for extended periods of time (Afifi *et al.*, 2012).

Third, in many instances, internal migration is a deeply embedded strategy for helping families cope with environmental changes affecting livelihoods in places of habitual residence. For example, studies examining the impact of drought on migration patterns in the Sahelian region of West Africa confirm that circular, internal migration is a traditional coping mechanism (Cordell and Piché, 1996; Afifi *et al.*, 2012). A study of displaced populations in the east and Horn of Africa, home to a significant number of traditional pastoralist communities, also noted that movement away from homelands was typically of an internal, circular and temporary nature (Afifi *et al.*, 2012: 12–13).

It is important to note, however, that internal and cross-border movement, while clearly demarcated by the existence of international borders, nevertheless can exist along a single mobility continuum. Recent studies confirm that, in a number of instances, while initial migration or displacement has been within states, secondary movement across borders has also occurred (Warner and Afifi, 2013). Recognising and understanding these complexities and the need for a comprehensive approach which helps states address both internal and cross-border population movements, Recital 22 references the work of the Nansen Initiative on disaster-induced, cross-border displacement. Established by the Swiss and Norwegian governments, the Nansen Initiative seeks to build a state-led protection agenda for cross-border displacement. Recital 23 records the clear intention of the drafters that the Peninsula Principles are to be regarded as complementary to the Nansen Initiative.

Recital 18 recognises the vast amount of work being undertaken by a multiplicity of researchers, academics, NGOs and United Nations or state agencies. As noted above, this body of work has been instrumental in increasing the empirical evidence as to the linkages between human mobility and climate change. This body of work not only increasingly documents the lived reality of affected individuals and communities across the globe, but has also contributed greatly to increasing the awareness of the issue among policy-makers at the national, regional and international levels.

However, no universally accepted definition of those who move due to the impacts of climate change has emerged. The early references in the literature to 'environmental refugees' are largely discredited as legally inaccurate, and proposals to amend the Refugee Convention are both politically infeasible and potentially destabilising to the whole refugee regime (Burson, 2010; Kälin and Schrepfer, 2012; McAdam, 2012; UNHCR, 2008). Beyond this broad consensus, the terminology and definitions used to describe persons forced to move due to environmental migration become more open to debate.

The definitions used in the Peninsula Principles are set out Principle 2. Reflecting the grounding of the Principles in an increasingly scientifically established reality of a changing global climate, the Principles align their definition of climate change with that used by the IPCC. Principle 2.a defines climate change as 'the alteration in the composition of the global atmosphere that is in addition to natural variability over comparable time periods'.

Principle 2.b defines climate displacement as being the movement of people due to the effects of climate change. The focus on the act of movement rather than its nature recognises that it will usually be difficult to categorise the movement as being wholly voluntary or forced. The definition does not seek to delineate where on the forced-voluntary spectrum any movement must exist for the Principles to operate. This is important as it means that the preparation and planning obligations outlined in Part II are not contingent on the existence of forced displacement, but rather arise immediately.

The definition also recognises the typical lack of a direct or mono-causal relationship between population movement and climatic factors. The definition, at Principle 2.b, makes clear that the fact that the effects of climate are mediated through a range of other factors such as age, gender, socio-economic status and perception of risk does not prevent the underlying cause being climate displacement. Importantly, 'climate-displaced persons', for the purposes of the norms and standards contained in the Principles, are defined to include not just individuals but also households and communities, in recognition that climate displacement has significant implications at both individual and group levels. Climate displacement affects families *qua* families and communities *qua* communities.

Another critical element of the definition of climate-displaced persons is that it includes not just those experiencing climate displacement but also those who are 'facing' it. This captures a key concern that steps taken to protect the lives and livelihoods of persons and groups impacted by the negative effects of climate change should not be left until after the event when the damage has been done. Rather an *ex-ante* approach is called for, which minimises the risk of future harm by undertaking preventive measures including planned and voluntary relocation to alternative locations.

While definitions matter in that they define boundaries of applicability of legal regimes – including the legal regime of the Peninsula Principles – it is nevertheless important not to get too distracted by debates on terminology. Although complex in nature, there is an increasingly recognised and understood relationship between environmental stressors and human mobility generally, and the provision of comprehensive protection to affected individuals and communities via a coherent legal regime which responds to the complexities involved is more important than the label used.

Theme 2: Protecting rights and framing responsibilities

The Peninsula Principles sit within and reflect an extensive and overlapping body of international law, standards and guidelines relating to human rights, internal displacement and natural disasters. One of the critical objectives of the Peninsula Principles is to articulate a coherent normative framework which, while drawing on these sources, contextualises and tailors this body of 'hard' and 'soft' law instruments to the particular challenges of climate change-related displacement within states.

Protecting rights

The Peninsula Principles recognise that climate-displaced persons are *already* rights-holders under existing general international human rights instruments. The Preamble references the Universal Declaration of Human Rights 1948, and the twin 1966 Covenants which give expression to the rights proclaimed therein: the International Covenant on Civil and Political Rights (ICCPR) and the International Covenant on Economic, Social and Cultural Rights (ICESCR). Collectively, these instruments are known as the International Bill of Rights and are referenced in Recital 2. Climate change will indubitably impact on a range of rights across these instruments.

As already noted, in many instances those most vulnerable to the disaster impacts of climate change are from minority and politically disfranchised communities living on marginal land. The non-enjoyment of basic civil and political rights under the ICCPR, such as the right to receive and impart information, to hold opinions, and to associate, gives rise to a lack of voice, and increases disaster risk by preventing communities vulnerable to natural hazards from any meaningful participation in disaster risk-management planning and risk-reduction programmes (IUCN, 2008). However, the impacts of natural disasters and climate change will resonate most loudly in the economic, social and cultural sphere. Therefore a rights-based approach that encompasses and promotes the enjoyment of participatory rights will also facilitate the enjoyment of socio-economic and cultural rights. For this reason, Recital 2 references the Vienna Declaration and Programme of Action, which recognise that human rights are held not just universally, but indivisibly as well.

Regarding the impact on the enjoyment of socio-economic rights, the effects of climate change on the environment will have significant impacts on the right to an adequate standard of living of many millions of people. A recent report on sectoral impacts on the Pacific has highlighted negative impacts of climate change on tourism and fishing, both important sectors for regional livelihood generation (Asian Development Bank, 2013).

The destruction of homes and assets already impacts on the right to adequate housing under Article 11 of the ICESCR. Already, it is estimated that some 150 million people are living in cities with significant water shortages (Foresight, 2011: 110). A rise in surface temperature is projected to lead to increased water stress impacting upon both the right to adequate water as a component of the right to an adequate standard of living, and the right to the highest standard of health under articles 11 and 12 of the ICESCR.[4] Rising temperatures are likely also to result in declining crop yields in many places. More recent studies over the last decade suggest a more rapid decline in crop yields due to global warming than previously thought; and increases in temperature may, in some countries, be off-setting production gains resulting from technological advances and other factors (World Bank, 2012). The destruction of arable land by wind and coastal erosion or by salt water intrusion arising from more intense and frequent storm surges will exacerbate stresses on the availability of land and water to meet both national and global demand for food (Food and Agriculture Organisation, 2011). These trends will impact on the right to adequate food under Article 11 of the ICESCR. Respiratory illness from rising temperatures, exposure to illness due to changes in disease vectors, and increases in emotional and psychological trauma linked to changes in the physical environment will impact upon the right to the highest standard of health (World Bank, 2012).

Disaster effects, including displacement, can cause significant and long-lasting psychological harm, which can also substantially impact on the right to health. As noted in the SREX report, while often overshadowed by the post-disaster physical health outcomes, 'the psychological effects can be long lasting and can affect a large portion of a population'. Common outcomes include anxiety and depression and indirect impacts could arise during the recovery period associated with the stress and challenges of loss, disruption and displacement (Handmer *et al.*, 2012: 252).

Significantly, as recognised in Recital 1, losing land or being forced to move away from ancestral land raises significant and profound issues of cultural loss. This is a particularly acute issue for indigenous communities whose traditional lands are in many instances under significant threat from the impacts of climate change. A number of studies on current and projected impacts of climate change on areas inhabited by indigenous and traditional peoples have been undertaken in recent years (IUCN, 2008; Afifi *et al.*, 2012). They note that climate change is already having a serious impact upon these peoples' livelihoods and culture. The affected communities have had to engage in adaptation measures including changes to traditional farming techniques and hunting and gathering periods and habits, rainwater harvesting, and crop and livelihood diversification.

Climate change raises complex issues around the rights of 'peoples' to self-determination, guaranteed under Article 1 of both the ICCPR and the ICESCR. While the issues raised are less complex in an internal displacement setting, nevertheless the movement of a people away from lands which they may be viscerally and existentially connected to onto other land comprising the traditional territory of other peoples has serious implications for the preservation of culture, identity and the right to freely pursue cultural development provided under Article 1 of the ICESCR.

Framing national responsibility

The Principles draw on the International Bill of Rights as the key international human rights instrument creating binding obligations on states to respect, protect and fulfil the rights of all persons within their jurisdiction on a non-discriminatory basis.

Under this body of international law, states bear the primary responsibility for protecting human rights. As the foregoing discussion makes clear, this general obligation includes protecting people from the foreseeable risks arising from natural disasters, including those arising through events and processes linked to climate change. Decisions from the European Court of Human Rights in *Öneryildiz* v. *Turkey*[5] and *Budeyeva et al.* v. *Russia*[6] establish that this responsibility includes taking steps to avoid loss of life and property from known environmental hazards. This core Principle of international human rights law is expressly acknowledged in Recital 11.

One key development in international human rights law has been the development of thematic human rights treaties which oblige states parties to be sensitive to the specific needs of vulnerable groups in society such as women, children and indigenous peoples. This is important because mobility-related adaptive capacity is variable, with vulnerable sections of society often disadvantaged. One study of a number of indigenous groups found capacity and vulnerability

> can be unevenly distributed between tribes and can even be asymmetrically distributed within a community. Women are particularly affected by climate change as a result of their disproportionate involvement in reproduction work, insecure property rights, limited access to resources, and reduced mobility.
>
> (IUCN, 2008: 5)

While there are a number of second-generation international human rights treaties dealing with matters of thematic concern such as racial discrimination (the Convention on the Elimination of Racial Discrimination, 1965), discrimination against women (the Convention on the Elimination of All Forms of Discrimination Against Women, 1979) and the rights of children (the Convention on the Rights of the Child, 1989), there is no universally applicable treaty dealing with the protection of those who are displaced internally.

However, Recital 3 explicitly references the UN Guiding Principles on Internal Displacement ('the Guiding Principles') and declares the intention of the Peninsula Principles to 'build on and contextualise' this normative framework. Although a soft law instrument and technically non-binding on states, the Guiding Principles have proved influential in setting policy and are a source of inspiration behind the African Union Convention for the Protection and Assistance of Internally Displaced Persons in Africa (the Kampala Convention), which is referred to in Recital 21.

The Peninsula Principles build upon and contextualise the Guiding Principles in the specific context of climate change in two ways. First, by removing any ambiguity as to the scope of application of the norms contained in the Guiding Principles to all who move to adapt to or avoid the adverse impacts of environmental events and processes linked to climate change. There is no doubt that, by including in its definition reference to natural or manmade disasters, the Guiding Principles will draw into its orbit many persons affected by climate change, particularly those at the forced end of the spectrum. However, the Guiding Principles were drafted at a time when the specific forms and contexts of forced migration and displacement were less well understood. There is some uncertainty as to the extent to which the definition of a displaced person contained in the Guiding Principles encompasses persons who move to avoid becoming a disaster victim *in the future* (Bronen, 2014; Koser, 2008). Some movement in anticipation of the worst, particularly in relation to slow-onset processes, will constitute a form of voluntary adaptive migration.

Yet, in many cases, the pre-emptive movement of people to avoid the worst impacts of climate change will feature some degree of compulsion. Faced with more frequent and intense sudden-onset events, an otherwise natural hazard-resilient household or community may over time change into a vulnerable one. Perceptions of risk may also fluctuate over time in response to changing environmental conditions. Drawing on the increased understanding which has arisen since the drafting of the Guiding Principles, the Peninsula Principles seek to *expressly close* any normative gap by expressly including in its definition of climate-displaced persons not just those experiencing climate displacement but also those *facing* it.

Also omitted from the scope of the Guiding Principles are internal labour migrants, compelled to leave their homes or places of habitual residence to avoid poverty (Koser, 2011: 293). While those whose motivation for movement is purely economic would also fall outside the scope of the Peninsula Principles, in many cases economic and environmental drivers mingle. There may well be a significant element of economic compulsion to move elsewhere in order to generate income to assist with *in situ* adaptation or otherwise promote resilience, particularly vis-à-vis slow-onset processes linked to climate change. For example, it is estimated that, between 1999 and 2001, more than 250,000 people, comprising approximately 20 per cent of the total regional population, migrated from Karakalpakstan in western Uzbekistan to Kazakhstan or the Russian Federation as a result of the combined effect of environmental degradation and unemployment (Glantz, 2005). The Peninsula Principles seek to further close any normative gap by including within their scope persons whose displacement occurs by reason of environmental factors in concert with other factors, including economic ones.

The second way in which the Peninsula Principles build on and contextualise the Guiding Principles is to align their normative content with the more programmatic and disaster-specific focus of the Hyogo Framework of Action 2005–2015[7] and other soft law instruments such as the UN Principles on Housing and Property Restitution for Refugees and Displaced Persons and the Inter-Agency Standing Committee (IASC) Operational Guidelines on the Protection of Persons in Situations of Natural Disasters.

These other soft law instruments are referenced in Recital 20 of the Preamble.

The Hyogo Framework of Action is particularly important in this context. With over 160 states present at its adoption, it recognises both that states bear the primary responsibility for protecting their citizens, and that developing states are more vulnerable to natural disasters. It sets out a range of priority actions with the overall aim of promoting community resilience and reducing disaster losses (paras 14–20). Many of the priority actions outlined in the Hyogo Framework of Action are reflected and adapted to the climate displacement context in the substantive text of the Peninsula Principles. Thus, Principle 9 deals with displacement risk management, Principle 10 outlines detailed steps in relation to community participation and consent, Principle 11 deals with steps relating to land identification and habitability, and Principle 13 contextualises the Hyogo Framework's emphasis on the development of appropriate legislative and policy measures, and the strengthening of national and local capacity. Principle 3 augments this programmatic regime by drawing on the fundamental human rights principle of non-discrimination, and the rights and freedoms of those who suffer displacement. By this amalgamation, the Peninsula Principles aim to provide a comprehensive regime encompassing the normative and operational elements required to protect the rights of climate-displaced persons.

Framing international responsibility

It will be difficult to establish a direct causality between an emitting state's actions and any particular instance of displacement, thus engaging legal obligations under international law (Kälin and Schrepfer, 2012). Such claims will likely fail. For this reason, a strict claim-based approach to the question of international 'responsibility' may not be advantageous. Nor, however, is it feasible to suggest that states facing or experiencing climate displacement should be left to fend for themselves. Responsibility for protecting the rights of those displaced by climate change cannot simply be considered the sole responsibility of the domestic state, it also has an international component. The Peninsula Principles therefore conceptualise the issue of climate displacement as a matter of global responsibility. This international dimension is reflected in Recital 12, which records the drafters' belief that, given the global dimension of climate change, states should, upon request by an affected state, provide adequate and appropriate support not only for mitigation and adaptation measures, but also in respect of 'relocation and protection measures, and provide assistance to climate displaced persons'.

There are two imperatives which drive such a position. First, there is a clear moral imperative in that those most at risk of losing enjoyment of rights are persons with basic or even subsistence-level lifestyles, who are citizens of countries with relatively low or modest levels of economic development. As such, they bear little responsibility for the build-up of dangerous levels of atmospheric greenhouse gases (GHGs) driving climate change.

Second, there is a practical imperative. Less developed nations have a higher proportion of their population living in low-lying coastal areas than developed nations (McGranahan *et al.*, 2007: 26). Yet developing nations will have relatively fewer resources to undertake mitigation activity, or to enable vulnerable populations living in areas exposed to natural hazards to adapt to the negative impacts of climate change. Also, state fragility and social vulnerability are correlated (Corendea *et al.*, 2012). Fragile states may lack the ability or political will to deliver goods and services to vulnerable groups to help them cope with environmental stressors. For example, weak governance arising from decades of conflict in Somalia increased vulnerability to drought conditions during 2010–2011 and contributed to significant levels of displacement (Lindley, 2014). Even when the political will is present, fragile states may simply lack the institutional capacity, and prioritise short-term interventions that do little to address vulnerability to repeated events. Translating this into a human rights perspective, developing states may not have the technical, personnel or financial capacity to take steps to discharge their obligations to protect human rights where the enjoyment of rights is threatened due to environmental stressors linked to climate change. This is reflected in Recital 11 which recognises that addressing and responding to climate displacement presents states with financial, logistical, political, resource and other difficulties.

Recital 13 observes that the international community has humanitarian, social, cultural, financial and security interests in addressing the problem of climate displacement in a timely, coordinated and targeted manner. Since the 1970s, increasing attention has been given to the linkage between environmental issues and both armed conflict and security. Although this argument is controversial at times, there is a general acceptance that environmental issues can pose threats to security and induce violent conflict and migration, albeit in a highly uncertain manner and through complex social and political processes (Baechler, 1999; Gleditsch, 1998, 2012; Ronnfelt, 1997). If significant numbers of persons are compelled to flee their homes and occupy land owned by others, perhaps for significant amounts of time, or consume scarce resources traditionally enjoyed by other groups, this can be a source of friction. Recital 10 recognises that climate displacement, if not properly managed, can be the cause of tension and instability within states. For example, it has been recently argued that the movement of coastally dwelling Bangladeshis into India via the Sunderbans region in response to flooding and coastal erosion may add to tension between the two countries (Bose, 2014). It has also been reported that the influx of lowland Bengali settlers into hill tracts has led to tension between the Chittagong hill tribes, who traditionally inhabited the area, and the government (Gleditsch *et al.*, 2007: 5). The movement of pastoralists in

Ethiopia to other regions of the country has also reportedly created conflict with other groups (Corendea *et al.*, 2012: 21).

In emphasising an international dimension to the responsibility for protecting the rights of climate-displaced persons, the Peninsula Principles draw on the duty to cooperate under international law. Recital 2 refers to the Charter of the United Nations ('the UN Charter'), the foundation stone of the modern architecture of international relations. Its provisions make clear that responsibility for the promotion and protection of international human rights rests with member states and with the international community. Indeed Article 1(3) of the UN Charter proclaims one of the purposes of the United Nations to be:

> To achieve international co-operation in solving international problems of an economic, social, cultural, or humanitarian character, and in promoting and encouraging respect for human rights and for fundamental freedoms for all without distinction as to race, sex, language, or religion.

This purpose is given concrete expression in Chapter IX, which contains a number of articles dealing with international economic and social cooperation. Article 55 of the Charter provides that the United Nations 'shall promote' a number of specified matters including 'higher standards of living, full employment, and conditions of economic and social progress and development' and 'solutions of international economic, social, health, and related problems'. Also specified is 'universal respect for, and observance of, human rights and fundamental freedoms for all'. The Principle is further embedded in Article 56 of the UN Charter, which obliges states to act jointly and separately for the realisation of human rights as well as economic and social progress and development. The duty to cooperate can therefore be regarded as an elemental component of the modern system in international relations.

The centrality of the duty on states to cooperate to promote and protect human rights is given particular emphasis in Article 2(1) of the ICESCR. The duty to cooperate is emphasised in the work of various treaty-monitoring bodies. In its General Comment No. 3: the Nature of States Parties Obligations under the Covenant, the Committee on Economic, Social and Cultural Rights ('the ESCR Committee') draws upon articles 55 and 56 of the UN Charter to emphasise that 'international cooperation for development and thus for the realisation of economic social and cultural rights is an obligation of all states'.[8] In General Comment No. 12: The Right to Adequate Food, the ESCR Committee again emphasised the 'essential role' of international cooperation in accordance with the UN Charter and the relevant articles of the ICESCR to facilitate access to food and to provide necessary aid.[9]

The ESCR Committee has expanded on its understanding of the normative content of the duty to cooperate to promote enjoyment of human rights in other general comments dealing with specific ICESCR rights. In General Comment No. 4: The Right to Adequate Housing the Committee observed that Article 11(1) of the ICESCR concludes with the obligation on states parties to recognise 'the essential importance of international cooperation, based on free consent'.[10] It emphasised that international cooperation should be targeted at disadvantaged groups most at need, and that states, when requesting international assistance, should do so in full consultation with affected groups. In General Comment No. 14: The Right to the Highest Attainable Standard of Health the ESCR Committee was more explicit, asserting that the duty encompassed not just a duty on a state unable to progressively realise the particular right through its own resources to request international assistance, but also a corresponding duty on the state requested to provide it, where able to do so. The Committee stated that it was 'particularly incumbent upon those states and other actors in a position to assist to provide international assistance and cooperation'.[11]

The United Nations Convention on the Rights of the Child (CRC) 1989, at Article 4, similarly provides that states parties have an obligation to secure economic social and cultural rights of children 'within the framework of international cooperation'. The Committee on the Rights of the Child has also emphasised the duty on states to seek international assistance in its General Comment No. 5: General Measures of Implementation, where it stated that implementation of the Convention was 'a cooperative exercise of the states of the world'.[12]

As reflected in Principle 8.d, it is implicit in the notion of a shared responsibility that the state affected by the adverse impacts of climate change should not arbitrarily refuse an offer of assistance from other states, if it is unable to protect the rights of those facing or experiencing displacement within its national borders. As Salomon (2006: 104) remarks, 'this would reasonably entail an obligation to countenance the request – that is, to discuss, consider, and respond satisfactorily'. The existence of some duty on an affected state to accept outside assistance to provide humanitarian relief when it lacks capacity to do so is supported by General Assembly resolutions 45/100 (1990)[13] and 46/182 (1991).[14] The latter, in particular, deals with the content of this duty. While it affirms the right of the sovereign state to decide whether or not to accept offers of assistance, this is qualified by the statement that those states with populations in need of humanitarian assistance 'are called upon' to facilitate the work of humanitarian organisations to deliver essential aid including food, shelter and medical and health care.[15] The existence of such a duty is further supported by the finding of the International Court of Justice in *Nicaragua* v. *United States*[16] where the court held that: 'the provision of strictly humanitarian aid to persons or forces in another country ... cannot be regarded as unlawful intervention, or in any other way contrary to international law'.

The ESCR Committee has explicitly addressed the duty to cooperate in the context of populations displaced by natural disasters.[17] It stated:

> States have a joint and individual responsibility, in accordance with the Charter of the United Nations, to cooperate in providing disaster relief and humanitarian assistance in the times of emergency, including assistance to refugees and internally displaced persons. Each State should contribute to this task in accordance with its ability.

This aspect of the duty to cooperate has also been emphasised by the Office for the High Commissioner for Human Rights in its 2009 Report on the Relationship Between Climate Change and Human Rights.[18] The International Law Commission's Draft Articles on the Protection of Persons in the Event of Disasters also provide that states have the duty to seek international assistance where they lack domestic capacity to cope and that consent to external assistance should not be arbitrarily withheld.[19]

While the pronouncements of the various treaty-monitoring bodies are technically not binding on states under international law, given that these bodies have been tasked under the particular treaty in question with providing guidance on the interpretation of the convention, they are of persuasive authority. While not universally adopted, both the ICESCR and the CRC have been adopted widely by states, so that the exposition of the normative content of a duty to cooperate in the context of natural disasters by their treaty-monitoring bodies potentially has a far reach.

Nevertheless, it is important not to overstate matters. While Article 56 of the UN Charter has proved particularly influential in the development of norms relating to the protection of human rights, the scope of the duty to cooperate under international law is uncertain (Stoll, 2012). Even in the field of the protection of economic social and cultural rights where it has proved most influential, as Alston and Quinn (1987: 191) note, it is difficult, if not impossible, to characterise the content of the duty to cooperate under the ICESCR as extending to a specific obligation on 'any particular state to provide any particular form of assistance'.

Yet, despite its uncertain reach, the duty to cooperate under international law is significant precisely because many of the impacts of climate change will be felt in the sphere in which it is most developed. Given the emphasis on international cooperation in the UN Charter, it can be said that where climate displacement as defined in the Peninsula Principles occurs or is faced, the international community has a shared responsibility to cooperate to prevent such displacement leading to the erosion of rights.

International responsibility within the context of the UNFCCC

Recital 14 notes that the response by states to the human mobility implications of climate change has generally been ad hoc and uncoordinated. The inclusion of a reference to human mobility within the framework of processes linked to the United Nations Framework Convention on Climate Change (UNFCCC) is a welcome step in the right direction. Recitals 16, 17 and 18 reference the relevant processes. The lengthy and complex history behind attempts to have climate displacement included in key international documents relating to climate change such as the UNFCCC and the Kyoto Protocol is referred to in Recital 15.

Warner (2011) provides a useful analysis of the status of human mobility as an issue within the wider UNFCCC process. She notes how, in the run-up to Fifteenth Session of the Conference of the Parties (COP 15) in Copenhagen in 2009, the policy and research community drew the attention of policy-makers to the issue, and succeeded in having migration and displacement included in the outcome text as an aspect of adaptation measures which required steps to promote 'enhanced understanding, coordination, and cooperation'. The legal process generated at COP 15 reached fruition at the Sixteenth Session of the Conference of the Parties (COP 16) in Cancún, when parties adopted the Cancún Adaptation Framework, Article 14(f) of which:

> 14. Invites all Parties to enhance action on adaptation under the Cancun Adaptation Framework, taking into account their common but differentiated responsibilities and respective capabilities, and specific national and regional development priorities, objectives and circumstances, by undertaking, inter alia, the following:
>
> ...
>
> (f) Measures to enhance understanding, coordination and cooperation with regard to climate change induced displacement, migration and planned relocation, where appropriate, at national, regional and international levels;

As Warner (2011: 12) notes, the recognition that different forms of human mobility may arise in the context of climate change is important. Earlier UNFCCC texts had referred to migration and displacement, and at one point had even referred to 'climate refugees'. By reflecting a more nuanced and considered understanding of the human mobility implications, Article 14(f) provides a solid foundation for national and international-level action for protecting the rights of those who have to move due to the negative effects of climate change.

Theme 3: Planned ex-ante relocation as the antidote to displacement

Recital 1 of the Preamble records the drafters' basic concern that displacement arising from events and processes caused or exacerbated by climate change can have significant negative impacts on the enjoyment of human rights. At the very minimum, it can lead to loss of housing, land and property rights, destroy livelihoods and, in some instances, at certain thresholds of climate change, lead to a loss of cultural identity.

For this reason, Recital 7 of the Preamble affirms the basic right of climate-displaced persons, as defined in the Principles, to remain in their places of residence for as long as possible through mitigation and adaptation measures. This right to remain is buttressed by Recital 4 which provides that 'when an activity raises threats of harm to human health, life or the environment, precautionary measures should be taken'.

Commenting in the specific context of sea-level rise and the residual uncertainty of its mobility impacts, Oliver-Smith (2011: 162) makes the point that 'it is clear that governments and international agencies would be very derelict in their responsibilities if they did not attempt to develop adequate responses to these questions.' Both Recital 4 and Oliver-Smith's plea draw on the precautionary Principle, an emerging Principle of international environmental law. Its most well-known articulation is to be found in Principle 15 of the 1992 Rio Declaration on Environment and Development. This states:

> In order to protect the environment, the precautionary approach shall be widely applied by States according to their capabilities. Where there are threats of serious or irreversible damage, lack of full scientific knowledge shall not be used as a reason for postponing cost-effective measures to prevent environmental degradation.

Yet climate change displacement is a reality and can be expected to become more so in the coming decades. While the institutional and policy responses needed to deal with existing and potential displacement in the future arising from sudden and slow-onset impacts of climate change are not identical, there is a common need for the issue to be addressed by policy-makers as an immediate or near-term policy issue in both cases (Boncour and Burson, 2010). While it may take years – if not decades – before the full impacts of long-term processes such as sea-level rise are known, the processes are already under way and it is essential that planning for these impacts begin now (Hugo, 2011).

Where movement away is or becomes unavoidable, Recital 8 affirms: 'The right of those who may be displaced to move safely and to relocate within their national borders over time'. The emphasis in Recital 8 on relocation of persons who may otherwise be at risk of displacement signals an important conceptual model underpinning the Peninsula Principles. In other words, the *ex-ante* internal relocation of individuals, households and communities vulnerable to the negative impacts of climate change is preferable to their *ex-post* displacement.

Insofar as Recital 8 affirms a *right* to relocation, it is grounded in international human rights law and, in particular, Article 12(1) of the ICCPR, which provides that 'everyone lawfully within a state shall, within that territory, have the right of liberty of movement and freedom to choose his residence.' While Article 12 would also cover displacement as a form of movement, the right to choose an alternative place of residence elsewhere in your country in anticipation of future displacement, a place which coincides with the choices made by other community members, falls within the scope of Article 12. While this is not an absolute right and the movement and choices of residence can be limited under Article 12(3), any limitations must comply with the traditional human rights criteria of being prescribed by law, being in pursuit of a legitimate aim and being no more than is necessary in the circumstances.[20]

The rights-oriented definition of relocation, referred to in Recital 8, reflects the acknowledgement in Recital 1 that current episodes of displacement often have significant negative impacts on the enjoyment of rights of those displaced. In this regard, many important lessons can be drawn from the development arena, as large-scale development projects have been the cause of significant levels of displacement, and have been the catalyst for the relocation of entire communities. Estimates of the numbers of persons displaced by large-scale development projects – typically dams – over the past 20 years are 280–300 million, with some 15 million being displaced annually (Ferris, 2012). Focusing on state-initiated relocation programmes, Hugo (2011: 268)[21] identifies a number of key features: such relocation has overwhelmingly been of an internal nature; it has predominantly involved rural-agricultural populations being resettled into rural-agricultural communities; the communities resettled have often been poor or marginalised, while the central government is typically of a strong nature; and consultation with the community being relocated has often been lacking and the relocation characterised by compulsion.

Regrettably, but perhaps unsurprisingly given these commonalities, it has been observed that the lessons learned from relocations in the wake of large-scale development projects are lessons of failure, not success (Cernea, 1997, 2000; Ferris, 2012). Examples of the type of issues that too often arise can be found in a study of experiences of relocated populations in Turkey arising from the construction of the Ataturk Dam in the southeastern part of the country, traditional home to Turkey's large minority Kurdish population as part of a large-scale, state-led development project. The construction of the dam flooded one entire district, and affected over 113,000 people from over 18,000 families located in 145 villages. Villagers could either receive compensation and self-relocate or be relocated by the state in an urban centre in the region or in a rural area outside the region (Parlak, 2007). Problems noted in the studies of affected populations include a lack of proper consultation and information, for example a lack of advice or guidance on how to invest compensation packages for confiscated land in circumstances where the persons had never received such a sum before, leading to capital wastage and impoverishment. Other problems included the break-up of extended family groups, unemployment and tensions with communities in places of relocation (Parlak, 2007; Kadirbeyoglu, 2010).

It is clear, then, that planned relocation, while a potential antidote to future displacement, also potentially carries significant downside risks for the enjoyment of basic human rights if undertaken poorly. This is reflected in Recital 9, which recognises 'the imperative to avoid such outcomes'. Yet successful relocation is not cheap and it is not quick. It takes money and time. Importantly, it also takes compassion and understanding of the particular needs and concerns of not only the community being relocated, but also the community into which they are to be relocated. Reflecting the imperative acknowledged in Recital 9, the Peninsula Principles, at Principles 10–16 contains a detail roadmap for states to ensure full participation and consent of both relocated and host communities, to give appropriate assistance to those relocated.

Recital 9 also recognises that relocation can have significant adverse impacts on the enjoyment of human rights. Drawing on his typology of proximate and non-proximate movement in the context of relocation (resettlement in his phraseology) in the Pacific region, Campbell observes (2010: 43) that the more one moves along the spectrum towards international movement, the higher the associated social, political, psychological and economic costs associated with the relocation. Even when planned relocation is undertaken with the full participation of an affected community, significant rights issues can arise. The resettlement in New Zealand in 1966 of approximately half of the Tokelauan population following a hurricane is a case in point. Longitudinal studies carried out on the Tokelauan community in New Zealand have shown that the resettled population and their children had a range of poor health outcomes relative to their kin at home (Lane *et al.*, 2005; Salmond *et al.*, 1985). This shows that relocation or resettlement of communities into new environments, even when leading to some form of durable solution, can inherently give rise to on-going protection needs.

Concluding remarks on the Preamble

It is clear that urgent meaningful mitigation action to reduce greenhouse gas emissions must be taken now by the international community if we are to avoid potentially catastrophic change in the planetary climate system over the coming decades. It is also increasingly clear that displacement within states will be one of many consequences of such change.

While the timeframes for climatic change stretch into decades, time is not on our side. Climate change-related displacement is already a reality to some extent. Given that we are already locked into a climate-affected future, it is vital that states start to plan now for the possibility that sections of their populations will not be able to remain living in their places of current residence, but will have to move elsewhere to preserve their lives and livelihoods. It is unclear how far technology and other adaptive measures will be able to shield *in situ* an increasing global population from the worst impacts. Some form of movement within the state may well become the new norm for increasing numbers of people.

The Preamble introduces the Peninsula Principles on Climate Displacement within States. It is hoped by the drafters that the Peninsula Principles, as contextualised by the three underlying themes contained within the Preamble, will act as a valuable guide in helping policy-makers and others frame the necessary steps to be taken to protect the rights of all persons who are forced to move in coming years and decades.

Notes

1 The flooding in 2011 was the worst for 50 years, and caused over 800 deaths in some 65 provinces. Approximately 14 million people were affected in total, and the floods caused damage and projected economic losses estimated at more than US$450 billion: UN Office for the Coordination of Humanitarian Affairs, *Thailand: Learning to Cope with Major Disasters*, 25 September 2012, www.ocha. org (accessed 15 January 2013).
2 The information, including citations in the following section, are largely taken from Handmer *et al.* (2012: 247–248).
3 See, for comparison, Article 31(2) of the Vienna Convention on the Law of Treaties 1969, United Nations, *Treaty Series*, vol. 1155, New York, p. 331.
4 See here General Comment No. 15 The Right to Water E/C.12/2002/11 (20 January 2003).
5 Application No. 48939/99 (30 November 2004).
6 Application Nos 15339/02, 21166/02, 20058/02, 111673/02 and 15343/02 (20 March 2008).
7 See Final Report of World Conference on Disaster Reduction A/Conf.206/6.
8 HRI/GEN/1/Rev.7. (2004) at paragraph 14.
9 E/C.12/1999/5 (1999) at paragraph 36.
10 E/1992/23 at paragraph 19.
11 E/C.12/2000/4 (2000) at paragraph 45.
12 CRC/GC/2003/5 (2003) at paragraph 60.
13 A/RES/45/100.
14 A/RES/46/182.
15 Ibid., at paragraph 6.
16 A/CN.4/590 at pp. 21–23.
17 E/C.12/1999/5 (1999) at paragraph 38.
18 A/HRC/10/61 (15 January 2009) at paragraph 99.
19 See Draft Articles 10 and 11(2).
20 For a summary of the understanding of the United Nations Human Rights Committee on the scope of Article 12, see General Comment No. 27: Article 12 (Freedom of Movement) CCPR/C/21/Rev.1/Add.9.
21 Hugo terms these 'resettlement'.

References

Afifi, T., Govil, R., Sakdapolrak, P. and Warner, K. (2012) *Climate Change, Vulnerability and Human Mobility: Perspectives of Refugees from the East and Horn of Africa*, Bonn, UNU–EHS.

Alston, P. and Quinn, G. (1987) The Nature and Scope of States Parties' Obligations under the International Covenant on Economic Social and Cultural Rights. *Human Rights Quarterly* 9: 157.

Asian Development Bank (2013) *The Economics of Climate Change in the Pacific*, Manila, ADB.

Baechler, G. (1999): Environmental Degradation in the South as a Cause of Armed Conflict. In A. Carius and K.M. Lietzmann (eds) *Environmental Change and Security*, Berlin, Springer, pp. 107–129.

Bohra-Mishra, P. and Massey, D.S. (2011) Environmental Migration and Out-migration: New Evidence from Nepal. In E. Piguet, A. Pécoud and P. de Guchteniere (eds) *Migration and Climate Change*, Cambridge, Cambridge University Press, pp. 74–101.

Boncour, P. and Burson, B. (2010) Climate Change and Migration in the South Pacific Region: Policy Perspectives. In B. Burson (ed.) *Climate Change and Migration: South Pacific Perspectives*, Wellington, Institute of Policy Studies.

Bose, S. (2014) Illegal Migration in the Indian Sunderbans. *Crisis Migration: Forced Migration Review* 45: 22.

Bronen, R. (2014) Choice and Necessity: Relocations in the Arctic and South Pacific. *Crisis Migration: Forced Migration Review* 45: 17–21.

Burson, B. (2010) Protecting the Rights of People Displaced by Climate Change: Global Issues and Regional Perspectives. In B. Burson (ed.) *Climate Change and Migration: South Pacific Perspectives*, Wellington, Institute of Policy Studies.

Burson, B. and Bedford, R. (2013) *Clusters and Hubs: Towards an Architecture for Voluntary Adaptive Migration in the Pacific*, Geneva, Nansen Initiative.

Campbell, J. (2010) Climate Change and Population Movement in Pacific Island Countries. In B. Burson (ed.) *Climate Change and Migration: South Pacific Perspectives*, Wellington, Institute of Policy Studies, pp. 29–42.

Carrington, D. (2013) Planet Likely to Warm by 4C by 2100, Scientists Warn. *Guardian*, 3 December 2013 (accessed 6 January 2014).

Cernea, M. (1997) The Risks and Reconstruction Model for Resettling Displaced Populations. *World Development* 25(10): 1569–1587.

Cernea, M. (2000) Risks, Safeguards and Reconstruction: A Model for Population Displacement and Resettlement. *Economic and Political Weekly* 35(3): 659–678.

Cordell, D., Gregory, J. and Piche, V. (1996) *How and Wage. A Social History of a Circular Migration System in West Africa*, Boulder, CO: Westview Press.

Corendea, C., Warner, K. and Yuzva, K. (2012) *Social Vulnerability and Adaptation in Fragile States*, Bonn UNU-EHS.

Dodman, D. and Satterthwaite, D. (2008) Institutional Capacity, Climate Change Adaptation and the Urban Poor. *Institute of Development Studies Bulletin* 39(4): 67–74.

Ferris, E. (2012) *Protection and Planned Relocation in the Context of Climate Change*, Geneva, UNHCR.

Food and Agriculture Organisation (2011) *The State of the World's Land and*

Water Resources for Food and Agriculture: Managing Systems at Risk, Rome, FAO.

Foresight: Migration and Global Environmental Change (2011) *Final Project Report*, London, The Government Office for Science.

Glantz, M. (2005) Water, Climate and Development Issues in the Amu Darya Basin. *Mitigation and Adaptation Strategies for Global Change* 10: 23–50.

Gleditsch, N.P. (1998) Armed Conflict and the Environment. *Journal of Peace Research* 35(3): 381–400.

Gleditsch, N.P. (2012) Whither the weather? Climate Change and Conflict. *Journal of Peace Research* 49(3).

Gleditsch, N.P., Nordås, R. and Salehyan, I. (2007) *Climate Change and Conflict: The Migration Link*, New York, International Peace Academy.

Handmer, J., Honda, Y., Kundzewicz, Z.W., Arnell, N., Benito, G., Hatfield, J., Mohamed, I.F., Peduzzi, P., Wu, S., Sherstyukov, B., Takahashi, K. and Yan, Z. (2012) Changes in Impacts of Climate Extremes: Human Systems and Ecosystems. In: C.B. Field, V. Barros, T.F. Stocker, D. Qin, D.J. Dokken, K.L. Ebi, M.D. Mastrandrea, K.J. Mach, G.-K. Plattner, S.K. Allen, M. Tignor and P.M. Midgley (eds) *Managing the Risks of Extreme Events and Disasters to Advance Climate Change Adaptation. A Special Report of Working Groups I and II of the Intergovernmental Panel on Climate Change* (IPCC). Cambridge, UK, and New York, Cambridge University Press, pp. 231–290.

Hugo, G. (2011) Lessons from Past Forced Resettlement for Climate Change Migration. In E. Piguet, A. Pécoud, and P. De Guchteneire (eds) *Migration and Climate Change*, Cambridge, Cambridge University Press, pp. 260–288.

Hugo, G. and Bardsley, D.K. (2014) Migration and Environmental Change in Asia. In E. Piguet and F. Laczko (eds) *People on the Move in a Changing Climate: The Regional Impact of Environmental Change on Migration*, Dordrecht, Springer, pp. 22–48.

Hirabayashi, Y. and Kanae, S. (2009) First Estimate of the Future Global Population at Risk of Flooding. *Hydrological Research Letters* 3: 6–9.

International Federation of the Red Cross (2014) Typhoon Haiyan – 100 Days on: Hopes and Fears for the Future, 17 February 2014, www.ifrc.org (accessed 27 February 2014).

IPCC (1992) *Climate Change: The IPCC 1990 and 1992 Assessments*, New York, IPCC.

IPCC (2012) *Managing the Risks of Extreme Events and Disasters to Advance Climate Change Adaptation*, ed. C.B. Field, V. Barros, T.F. Stocker, D. Qin, D.J. Dokken, K.L. Ebi, M.D. Mastrandrea, K.J. Mach, G.-K. Plattner, S.K. Allen, M. Tignor, and P.M. Midgley. *A Special Report of Working Groups I and II of the Intergovernmental Panel on Climate Change* (the SREX report), Cambridge University Press, Cambridge, UK, and New York, Cambridge University Press.

IPCC (2013) Summary for Policymakers. In *Climate Change 2013: The Physical Science Basis. Contribution of Working Group I to the Fifth Assessment Report of the Intergovernmental Panel on Climate Change*, T.F. Stocker, D. Qin, G.K. Plattner, M. Tignor, S.K. Allen, J. Boaschung, A. Nauels, Y. Xia, V. Bex and P.M. Midgley, Cambridge, UK, and New York, Cambridge University Press.

IPCC (2014) *Climate Change 2014: Impacts, Adaptation, and Vulnerability. Part A: Global and Sectoral Impacts. Contribution of Working Group II to the Fifth Assessment Report of the Intergovernmental Panel on Climate Change*, ed.

C.B. Field, V.R. Barros, T. Dokken, K.L. Mach, M.D.T.E. Bilir, M. Chaterjee, K.L. Ebi, Y.O. Estrada, A.C. Genova, B. Girma, E.S. Kissel, A.N. Levy, S. Mac-Cracken, P.R. Mastrandrea and L.L. White, Cambridge, UK, and New York, Cambridge University Press.

Kadirbeyoglu, Z. (2010) In the Land of Ostriches: Developmentalism, Environmental Degradation and Forced Migration in Turkey. In T. Afifi and J. Jäger (eds) *Environment, Forced Migration and Social Vulnerability*, Berlin, Springer Verlag.

Kälin, W. (2013) Changing Climates, Moving People: Distinguishing Voluntary and Forced Movements of People. In K. Warner, T. Afifi, W. Kälin, S. Leckie, B. Ferris, S.F. Martin and D. Wrathall (eds) *Changing Climate, Moving People: Framing Migration, Displacement and Planned Relocation*, Bonn, UNU-EHS Policy Brief No. 8, pp. 39–43.

Kälin, W. and Scherepfer, N. (2012) *Protecting People Crossing Borders in the Context of Climate Change: Normative Gaps and Possible Approaches*, Geneva, UNHCR Legal and Protection Policy Research Series.

Kleinen, T. and Petschel-Held, G. (2007) Integrated Assessment of Changes in Flooding Probabilities due to Climate Change. *Climatic Change* 81(3): 283–312.

Kniveton, D., Schmidt-Verkerk, K., Smith, C. and Black, R. (2008) *Climate Change and Migration: Improving Methodologies to Estimate Flows*. Geneva, International Organisation for Migration, pp. 331–358.

Koser, K. (2008) Gaps in IDP Protection. *Forced Migration Review* 31: 17.

Koser, K. (2011) Climate Change and Internal Displacement: Challenges to the Normative Framework. In E. Piguet, A. Pécoud and P. de Guchteniere (eds) *Migration and Climate Change*. Cambridge, UK, Cambridge University Press, pp. 289–306.

Lane, J., Siebers, R., Pene, G., Howden-Chapman, P. and Crane, R. (2005) Tokelau: A Unique Low Allergen Environment at Sea Level. *Clinical and Experimental Allergy* 35: 479–482.

Leckie, S., (2013) *Finding Land Solutions to Climate Change: A Challenge Like Few Others*, Geneva, Displacement Solutions.

Leighton, M. (2011) Drought, Desertification and Migration. In E. Piguet, A. Pécoud and P. de Guchteniere (eds) *Migration and Climate Change*, Cambridge, UK, Cambridge University Press, pp. 160–187.

Lindley, A. (2014) Questioning 'Drought Displacement'. Environment, Politics and Migration in Somalia. *Crisis Migration: Forced Migration Review* 45: 39–43.

McAdam, J. (2012) *Climate Change, Forced Migration and International Law*, Oxford, Oxford University Press.

McGranaghan, G., Balk, D. and Addeerson, B. (2007) The Rising Tide: Assessing the Risks of Climate Change and Human Settlements in Low Coastal Elevation Zones. *Environment and Urbanisation* 19(1): 17–37.

Maloney, S.K. and Forbes, C.F. (2011) What Effect Will a Few Degrees of Climate Change have on Human Heat Balance? Implications for Human Activity. *International Journal of Biometeorology* 55(2): 147–160.

Oliver-Smith, A. (2011) Sea Level Rise, Local Vulnerability and Involuntary Migration. In E. Piguet, A. Pécoud and P. de Guchteniere (eds) *Migration and Climate Change*, Cambridge, UK, Cambridge University Press, pp. 160–187.

Parlak, G. (2007) *Dance of Life with Water: Dams and Sustainable Development*, Ankara, Turhan Kitabevi.

Peduzzi, P., Chatenoux, B., Dao, H., Herold, C. and Giuliani, G. (2011) *Preview Global Risk Data Platform*, Geneva, UNEP/GRID and UNISDR, preview.grid. unep.ch/index.php?preview=tools&cat=1&lang=eng.

Piguet, E. and Laczko, F. (2014) Regional Perspectives on Migration, the Environment and Climate Change. In E. Piguet and F. Laczko (eds) *People on the Move in a Changing Climate: The Regional Impact of Environmental Change on Migration*, Dordrecht, Springer, pp. 1–21.

Roberts, D.W. (2010) *Social Dimensions of Climate Change in Urban China*, Manila, Asian Development Bank.

Ronnfelt, C.F. (1997) Three Generations of Environment and Security Research. *Journal of Peace Research* 34(4): 473–482.

Salmond, C.E., Joseph, J.G., Prior, I.A., Stanley, D.S. and Wissen, A.F. (1985) Longitudinal Analysis of The Relationship Between Blood Pressure And Migration: The Tokelau Island Migrant Study. *American Journal of Epidemiology.* 122 (2): 291–301.

Salomon, M., (2006) Human Rights Obligations: Obstacles and the Demands of Global Justice. In B. Andreassen and S.P. Marks (eds) *Development as a Human Right* Cambridge, MA, Cambridge University Press, pp. 96–118.

Sherwood, S.C., Bony, S. and DuFresne, J.-I. (2014) Spread in Model Sensitivity Traced to Atmospheric Convective Mixing. *Nature* 505: 37–42.

Stoll, T. (2012) Article 56. In B. Simma, D.E. Khan, G. Nolte and A. Paulus (eds) *The Charter of the United Nations: A Commentary* (3rd edn), Oxford, Oxford University Press.

UN-HABITAT (2008) *State of the World's Cities 2008/2009: Harmonious Cities.* London, Earthscan.

UNHCR (2008) *Climate Change, Natural Disasters and Human Displacement: A UNHCR Perspective*, Geneva, UNHCR.

UNISDR (2011) *Global Assessment Report on Disaster Risk Reduction: Revealing Risk, Redefining Development*, Geneva, United Nations International Strategy for Disaster Reduction Secretariat, and Oxford, Information Press.

Warner, K. (2012) *Where the Rain Falls: Climate Change Food and Livelihood Security. Global Policy Report of Where the Rain Falls Project*, Bonn, United Nations University and CARE International.

Warner, K. (2011) *Climate Change Induced Displacement: Adaptation policy in the Context of the UNFCCC Climate Negotiations*, Geneva, UNHCR.

Warner, K. and Afifi, T. (2013) Human Migration: Emerging Patterns and Understanding. In K. Warner, T. Afifi, W. Kälin, S. Leckie, B. Ferris, S.F. Martin and D. Wrathall (eds) *Changing Climate, Moving People: Framing Migration, Displacement and Planned Relocation*, Bonn, UNU-EHS Policy Brief No. 8, pp. 39–43.

Warner, K., Hamza, M., Oliver-Smith, O., Renaud. F. and Julca, A. (2010) Climate change, environmental degradation and migration. *Natural Hazards (Special Volume): Extreme Events, Vulnerability, Environment and Society* 55(3).

World Bank (2012) *Turn Down the Heat: Why a 4 Degree Centigrade World Must Be Avoided*, Washington DC, World Bank.

5   General obligations

Bonnie Docherty

Introduction

The General Obligations of the Peninsula Principles establish the first duties of the instrument. The Peninsula Principles open with a Preamble, which is non-binding, and an Introduction, which guides interpretation of the document as a whole.[1] The General Obligations, by contrast, require action by states confronting displacement due to climate change. Principles 5 through 8 deal, respectively, with four areas:

- prevention and avoidance;
- provision of adaptation assistance, protection, and other measures;
- national implementation measures; and
- international cooperation and assistance.

The General Obligations build on precedent in human rights and international environmental law and exemplify the interdisciplinary nature of the Peninsula Principles. They have two main goals: to preempt climate displacement and to ensure full implementation of the Peninsula Principles.

The four obligations can be divided into pairs, which reflect the section's dual aims. Principles 5 and 6 make clear that the Peninsula Principles prioritize obviating the need for people to flee climate change. Principle 5 takes a legal approach, requiring states to comply with international law in order to "prevent and avoid conditions that might lead to climate displacement." Principle 6 calls for more concrete steps, including the provision of adaptation assistance and protection so that people can remain in their homes for as long as possible. Principles 7 and 8 seek to guarantee that all of the Peninsula Principles are operationalized. Principle 7 mandates that affected countries promulgate laws and create structures at the domestic level in order to facilitate implementation of the Principles. Principle 8 requires the international community to support national efforts with cooperation and assistance.[2]

The General Obligations draw on binding treaties and normative frameworks of both human rights and international environmental law. The 1998 Guiding Principles on Internal Displacement, which require states to protect specified human rights in the context of forced migration, are particularly germane because they apply to people displaced within national boundaries as a result, *inter alia*, of natural disasters.[3] The influence of that document and others related to displacement is evident in many elements of Peninsula Principles 5 through 8. Additional instruments of human rights law, notably the two core covenants and several treaties dedicated to specific categories of people, further underpin the obligations. Their use as standards reflects the value the Peninsula Principles place on human rights in general. The 1992 UN Framework Convention on Climate Change (UNFCCC), the most relevant international environmental law treaty, informs Principle 5, which treats climate change mitigation as a tool for minimizing displacement, and Principle 8, which mandates international cooperation and assistance.[4] The precedent underlying the Peninsula Principles' General Obligations grounds the four Principles firmly in international law.

This chapter provides an in-depth analysis of Principles 5 through 8. It identifies the sources of the Principles, which both buttress the provisions and provide clues to their interpretation. The chapter parses the language of the Principles to elucidate their requirements and show how specific obligations advance each Principle's aims. In many cases, it also gives examples of how states can implement a Principle. By enhancing under-standing of the General Obligations, this chapter illuminates their signifi-cance for minimizing the human cost of climate displacement.

Principle 5: Prevention and avoidance

> States should, in all circumstances, comply in full with their obliga-tions under international law so as to prevent and avoid conditions that might lead to climate displacement.

This Principle aims to preempt climate displacement before it happens. By making "prevention and avoidance" the first general obligation, the Pen-insula Principles emphasize that averting climate displacement should be a primary goal. In this context, prevention can be understood as a positive obligation requiring states to adopt proactive measures to obviate climate displacement. Avoidance can be interpreted as a negative obligation that prohibits states from acting in ways that could lead to the problem.

Principle 5 draws heavily on the Guiding Principles on Internal Dis-placement. This instrument, submitted to the UN Commission on Human Rights by the Secretary-General's representative on internally displaced persons, "constitute[s] a comprehensive minimum standard for the treat-ment of [internally displaced persons] and [is] being applied by a growing number of states and institutions."[5] The language of Principle 5 parallels that of the Guiding Principles, which state, "All authorities and inter-national actors shall respect and ensure respect for *their obligations under international law*, including human rights and humanitarian law, *in all circumstances, so as to prevent and avoid conditions that might lead to displacement* of persons" (language common to both italicized).[6] Principle 5 thus adapts widely accepted norms to the issue of climate displacement.[7]

Principle 5 takes a precautionary approach to prevention and avoidance. It does not target climate displacement specifically but instead calls for prevention and avoidance even of "the conditions that might lead to climate displacement." Use of the word "might" means that there need not be proof that climate displacement will definitely occur before states should take action. This approach is consistent with the precautionary Principle of environmental law. As articulated in the 1992 Rio Declaration on Environment and Development, which establishes norms for sustainable development, the precautionary Principle states, "Where there are threats of serious or irreversible damage, lack of full scientific certainty should not be used as a reason for postponing cost-effective measures to prevent environmental degradation."[8] The UNFCCC, the lead international treaty dedicated to combating climate change, also urges states to follow the precautionary Principle.[9]

Compliance with international environmental law helps states meet their obligation to prevent and avoid. Stopping or slowing climate change, the purview of environmental law, would help minimize conditions that might lead to climate displacement. Although the UNFCCC does not mention climate displacement, it calls on parties to "take precautionary measures to anticipate, prevent or minimize the causes of climate change and mitigate its adverse effects."[10] The treaty defines "adverse effects" as changes in the environment that "have significant deleterious effects ... on human health and welfare," which could include climate displacement.[11] The 2010 Cancún Adaptation Framework, agreed to by the UNFCCC's Sixteenth Conference of the Parties, more clearly expresses the need to address climate displacement, encouraging states to take "measures to enhance understanding, coordination and cooperation with regard to climate change induced displacement, migration and planned relocation."[12]

States that use, endorse, or otherwise adopt the Peninsula Principles should also abide by their specific duties under international human rights law that are of direct relevance to the question of climate displacement. The Guiding Principles on Internal Displacement highlight the applicability of human rights law to the prevention and avoidance of displacement in general, and this body of law should be taken into account when addressing the human impact of climate change.[13] Even if climate change continues unchecked, protecting human rights can reduce the chances that people will be compelled to flee their homes and lands in response.

States are obliged not to displace people arbitrarily. The Guiding Principles declare that "[e]very human being shall have the right to be protected against being arbitrarily displaced from his or her home or place of habitual residence."[14] They prohibit such displacement, *inter alia*, in "cases of disasters," which could include natural disasters caused by climate change.[15] According to Francis Deng, the representative of the UN Secretary-General on internally displaced persons, this right makes explicit an implicit obligation under international law.[16] It originates in the right to freedom of movement and the right to choose one's residence within a state, which appear in the 1966 International Covenant on Civil and Political Rights (ICCPR).[17] "[U]nless the safety and health of those affected requires their evacuation," therefore, a state may not forcibly remove persons using climate change as a justification.[18]

In the face of climate change, states should also uphold participatory rights, which seek to engage people in choices that affect them. Proper planning can help prevent conditions that might lead to climate displacement, and consultation with affected persons produces fairer and more effective plans. Some of the relevant rights of participation deal with the provision of information, involvement in decision making, and the prerequisite (except in cases of emergency) of free, prior, and informed consent before relocation. These rights and their sources will be discussed further below in the context of Principle 7d.

The obligations under Principle 5 apply broadly. They create duties both for states affected by climate change, which are required to protect the rights of their people, and for states that cause climate change because they can influence its mitigation and thus decrease the need for displacement. Principle 5 requires every state to comply "in full" with its obligations and to do so "in all circumstances."

Principle 6: Provision of adaptation assistance, protection and other measures

a   States should provide adaptation assistance, protection and other measures to ensure that individuals, households and communities can remain in their homes or places of habitual residence for as long as possible in a manner fully consistent with their rights.

b   States should, in particular, ensure protection against climate displacement and demonstrate sensitivity to those individuals, households and communities within their territory who are particularly dependent on and/or attached to their land, including indigenous people and those reliant on customary rules relating to the use and allocation of land.

Principle 6 requires states that are dealing with the prospect of climate displacement to take concrete steps to minimize disruption and allow people to stay in their homes. It establishes duties toward all persons threatened by climate displacement as well as special responsibilities toward certain groups. These obligations fall on affected states, i.e., states within which climate displacement could occur; paragraph b makes that explicit when it refers to people "within their territory."

Principle 6a

The goal of Principle 6a is to reduce the risk that climate change will compel people to migrate. It says that states should "ensure that individuals, households and communities can remain in their homes or places of habitual residence for as long as possible." With the last phrase, Principle 6a recognizes that people might eventually have to move if the situation grows dire, but it strives to avert or at least delay such displacement.

The language of the Principle indicates the breadth of its application. It extends its protections to a wide range of affected people by employing the phrase "individuals, households and communities," which appears throughout the Peninsula Principles.[19] The Guiding Principles on Internal Displacement similarly look at internally displaced persons individually and collectively, describing them as "persons or groups of persons."[20] This expansive view of affected people, sometimes called "victims," reflects a growing trend in international law toward a broad definition. For example, the 2005 Basic Principles and Guidelines on the Right to a Remedy and Reparation, adopted by the UN General Assembly, state that the term "victims" encompasses not only the individuals harmed, but also "the immediate family or dependants of the direct victim and persons who have suffered harm in intervening to assist victims in distress or to prevent victimization."[21] The 2008 Convention on Cluster Munitions, a groundbreaking disarmament instrument that adopts a human rights approach to victim assistance, casts an even wider net and defines "victims" as including "those persons directly impacted by cluster munitions as well as their affected families and communities."[22] In following this precedent, the Peninsula Principles ensure that those facing climate displacement will receive assistance and protection regardless of whether they migrate alone or in groups.

Principle 6a recognizes that affected people may be displaced from different kinds of locations, including "homes or places of habitual residence." The Guiding Principles on Internal Displacement repeatedly use that phrase, including in the definition of internally displaced persons.[23] The words indicate that property ownership is not required to be a "climate displaced person," nor is living in a specific structure.[24] The Principle applies to those, such as indigenous peoples, whose tie is to the land more than to a building. According to the 2010 *Handbook for the Protection of Internally Displaced Persons*, which contains operational guidance based on global consultations, "The term 'homes or places of habitual residence' does not necessarily refer to a house or a building but can also designate land on which groups traditionally live or depend for their livelihoods, as in the case of nomads or pastoralists."[25] The language of Principle 6a thus seems to cover everyone threatened with climate displacement.

To keep people in their homes or places of habitual residence, Principle 6a requires states to "provide adaptation assistance, protection and other measures." Assistance and protection are central to international efforts to relieve the plight of refugees and internally displaced persons.[26] As defined by the *Handbook for the Protection of Internally Displaced Persons*, humanitarian assistance is "[a]id that seeks to save lives and alleviate the suffering of a crisis-affected population."[27] It can be divided into three categories: direct assistance, indirect assistance, and infrastructure support.[28] Principle 6a calls for the provision of adaptation assistance that encompasses all these types of humanitarian aid in order to help those at risk of displacement adjust to environmental changes without having to flee. Aid could come, for example, in the form of financial support or material help such as building barriers against flooding, reinforcing structures in advance of storms, or providing food, water, and medical supplies in cases of emergency.

Protection involves respecting, protecting, and fulfilling the rights of refugees and internally displaced persons.[29] It "encompasses all activities aimed at obtaining full respect for the rights of the individual in accordance with the letter and spirit of international human rights, refugee and humanitarian law."[30] Protection enhances assistance by ensuring that it is delivered in a way that respects human dignity and the Principle of non-discrimination. It also promotes civil and political as well as economic, social, and cultural rights.[31] The discussion of Principle 5 above identifies some of the rights particularly relevant to Principle 6a's goal of allowing climate-displaced persons to remain in their homes: the right not to be arbitrarily displaced and the right to participate in decision making.

The last phrase of Principle 6a reiterates the importance of the human rights of those threatened by climate change. It explains that states should provide assistance and protection so that people can remain in their homes "in a manner fully consistent with their rights."[32] The interdisciplinary principle thus uses a human rights framework to address the impacts of climate change, an issue of international environmental law.

Principle 6b

Principle 6b seeks to ensure that states pay extra attention to the needs of people especially vulnerable to the impacts of climate displacement. It again covers "individuals, households and communities," but here focuses on those "particularly dependent on and/or attached to their land." That phrase derives from a similar one in the Guiding Principles on Internal Displacement, which requires states "to protect against the displacement of ... groups with a special dependency on and attachment to their lands."[33] Such people will suffer more if they have to leave it, so ensuring that they are not neglected is important in the context of climate displacement.

Principle 6b of the Peninsula Principles specifically identifies indigenous people as members of that category because of their unique and legally recognized relationship to the land. The widely endorsed UN Declaration on the Rights of Indigenous Peoples, adopted by the UN General Assembly in 2007, highlights in its Preamble indigenous peoples' "rights to their land, territories, and resources."[34] In its text, it refers to indigenous peoples' "distinctive spiritual relationship with their traditionally owned or otherwise occupied and used lands, territories, waters and coastal seas."[35] The 1989 International Labour Organization (ILO) Convention No. 169 on Indigenous and Tribal Peoples, which sets additional standards for the protection of these groups, includes similar language. It requires governments to "respect the special importance for the cultures and spiritual values of the peoples concerned of their relationship with the lands or territories, or both as applicable, which they occupy or otherwise use."[36] The definition of indigenous peoples also reflects their tie to specific territory. ILO Convention No. 169 describes indigenous peoples as those

> in independent countries who are regarded as indigenous on account of their descent from the populations which inhabited the country, or a geographical region to which the country belongs, at the time of conquest or colonisation or the establishment of present state boundaries and who, irrespective of their legal status, retain some or all of their own social, economic, cultural and political institutions.[37]

Principle 6b does not limit its protections to indigenous peoples; it covers "those reliant on customary rules relating to the use and allocation of the land." Examples of those with close ties to the land can be drawn from the Guiding Principles on Internal Displacement, which require protection against displacement for indigenous peoples and "minorities, peasants, pastoralists and other groups with a special dependency on and attachment to their lands."[38]

Principle 6b sets forth two obligations toward all of these people. First, states must "ensure protection against climate displacement." In addition, as states deal with the prospect of climate change, they must "demonstrate sensitivity" to individuals, households, and communities with ties to the land.

Principle 7: National implementation measures

a   States should incorporate climate displacement prevention, assistance and protection provisions as set out in these Peninsula Principles into domestic law and policies, prioritising the prevention of displacement.

b   States should immediately establish and provide adequate resources for equitable, timely, independent and transparent procedures, institutions and mechanisms – at all levels of government (local, state and national) – to implement these Peninsula Principles and give effect to their provisions through specially earmarked budgetary allocations and other resources to facilitate that implementation.

c States should ensure that durable solutions to climate displacement are adequately addressed by legislation and other administrative measures.

d States should ensure the right of all individuals, households and communities to adequate, timely and effective participation in all stages of policy development and implementation of these Peninsula Principles, ensuring in particular such participation by indigenous peoples, women, the elderly, minorities, persons with disabilities, children, those living in poverty, and marginalized groups and people.

e All relevant legislation must be fully consistent with human rights laws and must in particular explicitly protect the rights of indigenous peoples, women, the elderly, minorities, persons with disabilities, children, those living in poverty, and marginalized groups and people.

While other Peninsula Principles require states to take specific actions, Principle 7 obliges states to establish a legal and structural framework with the goal of facilitating implementation of the rest of the instrument. The Peninsula Principles, like many of their sources, place primary responsibility for dealing with climate displacement on affected states. This approach, which comes from both human rights and international environmental law, reflects a concern for preserving national sovereignty. Applying this basic tenet of human rights law to displacement situations, the Guiding Principles on Internal Displacement, for example, assign national authorities the "primary duty and responsibility to provide protection and humanitarian assistance."[39] The 2011 Nansen Principles, which offer general recommendations for a new international framework on climate change and displacement, similarly declare that "[s]tates have a primary duty to protect their populations and give particular attention to the special needs of the people most vulnerable to and most affected by climate change and other environmental hazards."[40] The UNFCCC also recognizes the "Principle of sovereignty of States."[41] These provisions reflect the UN Charter's prohibition on outside intervention in domestic matters as well as the UN General Assembly's affirmation of "the sovereignty of affected States and their primary role in the initiation, organization, coordination, and implementation of humanitarian assistance within their respective territories."[42]

Not all international instruments include provisions on national implementation measures, but such provisions, more common in recent treaties, are valuable for ensuring that laws and structures are in place to achieve the goals of the instrument. The 1998 Aarhus Convention, a European treaty that lays out requirements for public participation in decisions affecting the environment, requires national implementation measures in the first of its General Provisions.[43] Some human rights treaties, such as the

1989 Convention on the Rights of the Child and the 2006 Convention on the Rights of Persons with Disabilities, also require states to adopt legislative or administrative implementation measures.[44] The Peninsula Principles have more detailed implementation requirements, but they build on these precedents. The inclusion of Principle 7 strengthens the document and will facilitate realization of the other Principles.

Principle 7a

This paragraph requires states to incorporate the Peninsula Principles' provisions into "domestic law and policies." Legislation is generally a stronger tool of implementation than policy because it is binding and more difficult to amend. Those qualities are especially important in this case because the Peninsula Principles, if adopted as designed, would be non-binding soft law.[45] Legislation also has value because it gives a responsibility the symbolic weight of law. Principle 7a does, however, allow for the adoption of policies when they are appropriate for administrative details or in cases in which legislation is challenging or slow to pass.

The paragraph boils the Peninsula Principles' provisions down to three categories: prevention, assistance, and protection. As discussed above, Principle 5 most clearly articulates the obligation to take steps to prevent climate displacement, although Principle 6 reinforces it. The duty to provide assistance and protection is woven throughout the document and goes beyond Principle 6a's call for assistance and protection to help people remain in their homes. Principle 7a requires implementation of all three types of actions, but like the previous two Principles, it emphasizes efforts to avoid climate displacement. Its final phrase obliges states to "prioriti[ze] the prevention of displacement."

Principle 7b

Principle 7b focuses on measures to implement the Peninsula Principles at the structural level. It mandates the establishment of "procedures, institutions and mechanisms," which are discussed more concretely in subsequent Principles. Procedures could include, for example, the process outlined in Principle 10 to give climate-displaced persons the information and opportunity to participate in decision making, or in Principle 11 to identify and reserve land for relocation. Institutions would likely be government bodies, such as those designed to distribute the post-displacement emergency aid described in Principle 14. Mechanisms could encompass systems to grant remedies and compensation to victims of human rights violations, as called for in Principle 16. Each of these types of structures is essential to operationalizing the Peninsula Principles, and states should adopt them "immediately," before climate displacement begins and becomes harder to manage.

Principle 7b requires not only the establishment but also the "adequate" resourcing of procedures, institutions, and mechanisms. In the context of humanitarian aid, "adequate" means available ("in sufficient quantity and quality"), accessible ("provided to all," "within safe reach," and "known to the beneficiaries"), acceptable ("respectful of … culture"), and adaptable (provided in flexible ways).[46] Principle 7b highlights the need for "specially earmarked budgetary allocations," which could not be transferred to fund other programs. The paragraph also refers to "other resources," which could include, *inter alia*, material support or personnel. Principle 7b helps guarantee that the structures set up to implement the Peninsula Principles have the capacity to make a difference.

This Principle specifies that the systems it mandates should meet certain standards to ensure their effectiveness. Equitability promotes fairness and non-discrimination, a core tenet of human rights, which is also addressed in Principle 3. Timeliness encourages states to take steps to prevent climate displacement or to plan for unavoidable displacement before an emergency situation arises. Independence helps ensure that the efforts to deal with climate displacement do not become corrupted or politicized. Finally, transparency facilitates the participation of climate-displaced persons, who should have a voice in decisions that affect them, as well as public monitoring, which can hold states accountable. The same requisites appear in a provision on national procedures, institutions, and mechanisms in the Pinheiro Principles on Housing and Property Restitution for Refugees and Displaced Persons, adopted by the UN Sub-Commission on the Promotion and Protection of Human Rights in 2005.[47]

States should implement Principle 7b at "all levels of government (local, state and national)." Different levels are appropriate for different purposes. For example, consulting with climate-displaced persons might be done more effectively at the local level. Those persons might feel more comfortable communicating in a familiar and accessible setting, and officials would have the context better to understand their needs and desires. It might make sense for assistance to be distributed locally or regionally rather than to try to coordinate the provision of aid from afar. By contrast, identifying relocation sites and facilitating returns could benefit from a national perspective because people could migrate to any part of the country.

Principle 7c

This provision requires states to address the need for "durable solutions to climate change" in their law or policy. Durable solutions are measures to deal with the long-term situation of displaced persons in situations when displacement cannot be prevented. Once a durable solution is achieved, the affected persons "no longer have any specific assistance and protection needs that are linked to their displacement and can enjoy their human rights without discrimination on account of their displacement."[48]

The term is widely used in documents that address forced displacement.[49] For example, the 1993 Vienna Declaration and Program of Action, adopted by the World Conference on Human Rights and endorsed by the UN General Assembly, calls for "achievement of durable solutions" for refugees and for "lasting solutions" for internally displaced persons.[50]

International law generally treats voluntary return to the place of origin as the preferred solution to displacement.[51] Human rights law explicitly recognizes the right of return for cross-border refugees and implicitly does for internally displaced persons.[52] According to an analysis of the Guiding Principles on Internal Displacement, "such a right [for internally displaced persons] can be deduced from the right to the liberty of movement and the right to choose one's residence."[53] The Guiding Principles assign authorities the duty to "establish conditions, as well as provide the means, which allow internally displaced persons to return voluntarily ... to their homes or places of habitual residence."[54] The Peninsula Principles also prioritize return, as evidenced in Principle 17, which establishes a framework for return.

Unlike most of Principle 7, however, paragraph c does not limit itself to implementation of the Peninsula Principles; it thus permits states to pursue other solutions when voluntary return is not feasible. These solutions include "integration at the location [affected persons] were displaced to, or resettlement to another part of the country."[55] The Guiding Principles on Internal Displacement provide precedent for these options and require authorities to ensure displaced persons can choose among them.[56] Allowing for integration and relocation is important in the climate displacement context because climate change has the potential to make certain areas permanently uninhabitable. For example, people would not be able to return to an island that has sunk below sea level or to farmland that has become a desert.[57]

When implementing the Peninsula Principles, a state is obligated to "ensure that durable solutions to climate displacement are adequately addressed by legislation and other administrative measures." As noted above, legislation is the strongest form of implementation, but administrative measures may sometimes be more appropriate or feasible. Implementation measures should not only facilitate the range of solutions, but also guarantee that climate-displaced persons have the freedom to select any of them, in accordance with human rights law. The "notion of free choice" is "[a]t the core" of the provision on solutions in the Guiding Principles on Internal Displacement.[58] Other guidelines and principles dealing with displacement also emphasize the importance of choice.[59]

Principle 7d

This paragraph obliges states to ensure that climate-displaced persons can exercise their right to participate "in all stages of policy development and implementation of these Peninsula Principles." Other Principles, notably Principle 10, lay out guidelines for participation in the preparation and planning for climate displacement, but Principle 7d is more overarching and also addresses the creation of the policies behind those plans.

Multiple documents that address environmental issues through a human rights framework provide precedent for Principle 7d's guarantee of participation. The Aarhus Convention, which builds on the Rio Declaration's Principle 10, seeks to promote the right to a healthy environment by ensuring "rights of access to information, public participation in decision-making, and access to justice in environmental matters."[60] The convention requires, for example, authorities to collect and disseminate information related to environmental matters. They must also facilitate the engagement of interested public parties; states should make draft laws and regulations available, provide an opportunity for public comment, and take public input into account.[61] The Inter-Agency Standing Committee, formed by the UN General Assembly to coordinate humanitarian assistance, has urged states to respect the right to participation when protecting persons affected by natural disasters. The standing committee's operational guidelines state that "[a]ffected persons should be informed and consulted on measures taken on their behalf and given the opportunity to take charge of their own affairs to the maximum extent and as early as possible."[62] Like Principle 7d, the guidelines call for participation "in the planning and implementation of the various stages of the disaster response."[63]

Other sources that address either climate change or displacement offer legal support for mandating opportunities for participation. The UNFCCC requires states parties to promote awareness of climate change and its effects by, for example, providing access to information on the topic and facilitating public participation when developing responses to the problem.[64] The Guiding Principles on Internal Displacement oblige states to share information on displacement and to seek "free and informed consent" for displacement from those affected.[65] They also call for "full participation of internally displaced persons in the planning and management of their return or resettlement and reintegration."[66] To comply with Principle 7d of the Peninsula Principles, states should grant people threatened or affected by climate displacement the rights articulated in such international instruments.

According to Principle 7d, the nature of the participation should be "adequate, timely and effective." The Aarhus Convention uses the same terms to describe the manner in which states should provide the public with information necessary to participate in environmental decision making.[67] Adequacy ensures that participation is of a sufficient level to have an impact on policy development and implementation. Timeliness guarantees that participation takes place when there is still time to have influence. Effectiveness means that public participants have the possibility of changing the outcome because their views are taken into account. States that adopt the Peninsula Principles should facilitate this type of participation through "all stages" of dealing with climate displacement.

Like Principle 6b, Principle 7d strives to protect the rights of certain groups in particular. Like other Principles, it covers individuals, households, and communities. At the same time, it requires states to make extra efforts to ensure participation by categories of people who have historically had fewer such opportunities. These groups include "indigenous peoples, women, the elderly, minorities, person with disabilities, children, those living in poverty, and marginalized groups and people." The participatory rights of many of these categories of people derive from rights enshrined in international instruments specifically dedicated to them.[68]

Principle 7e

This Principle declares that national legislation related to climate displacement "must be fully consistent with human rights laws."[69] It should be interpreted broadly to refer to all civil and political rights as well as economic, social, and cultural ones applicable to displacement.[70] These rights should encompass the rights to housing, land, and property, which are mentioned in several places in the Peninsula Principles.[71] The rights to freedom of movement and the choice of one's own residence are also important in situations of climate displacement. The Guiding Principles on Internal Displacement enumerate many other relevant rights, such as the rights to life, dignity, liberty, and security (Principle 8), the right to an adequate standard of living (Principle 18), and the rights to freedom of expression and association (Principle 22).[72] This list is not exhaustive.[73]

Principle 7(e) calls for special protections for the rights of certain categories of people. The list is the same as that in Principle 7d: "indigenous peoples, women, the elderly, minorities, persons with disabilities, children, those living in poverty, and marginalized groups and people." These people have historically experienced greater difficulty in exercising their rights, and the Peninsula Principles seek to ensure that they do not face further obstacles in the context of climate displacement.

Principle 8: International cooperation and assistance

a   Climate displacement is a matter of global responsibility, and States should cooperate in the provision of adaptation assistance (to the maximum of their available resources) and protection for climate displaced persons.

b   In fulfilling their obligations to prevent and respond to climate displacement within their territory, States have the right to seek cooperation and assistance from other States and relevant international agencies.

c States and relevant international agencies, either separately or together, should provide such cooperation and assistance to requesting States, in particular where the requesting State is unable to adequately prevent and respond to climate displacement.

d States that are otherwise unable to adequately prevent and respond to climate displacement should accept appropriate assistance and support from other States and relevant international agencies, whether made individually or collectively.

While Principle 7 lays out guidelines for how states should implement the Peninsula Principles, Principle 8 seeks to ensure they will have the means to do so. It obliges the international community to cooperate with and assist affected states. Precedent for such a provision appears in both human rights and international environmental law. For example, the 1951 Refugee Convention, which addresses displacement across national borders, declares that international cooperation is necessary to solve the global refugee problem.[74] The UNFCCC lays out detailed requirements for developed countries to help developing ones cope with the adverse effects of climate change.[75] The Nansen Principles state that "[w]hen national capacity is limited, regional frameworks and international cooperation should support action at [the] national level and contribute to building national capacity."[76]

Principle 8a

Principle 8a recognizes that while climate displacement involves movement within states, the problem is global in nature. Countries around the world, particularly developed ones, contribute to climate change through the release of greenhouse gases and other harmful activities. The environmental effects transcend national boundaries and may lead to climate displacement. Principle 8a, therefore, declares that "[c]limate displacement is a matter of global responsibility." This sentiment echoes the UNFCCC's Preamble, which "acknowledg[es] that the global nature of climate change calls for the widest possible cooperation by all countries and their participation in an effective and appropriate response."[77]

The second part of Principle 8a obliges all states to cooperate in providing adaptation assistance and protecting climate-displaced persons' rights. International assistance for adaptation can come in a variety of forms. States can offer, for example, financial support, in-kind aid, personnel, and/or technical advice. It can be provided directly or via a fund or central clearinghouse. The Principle limits responsibility for states "to the maximum of their available resources," but given the many possible forms of assistance, every state should be able to contribute in some way.[78]

States should also cooperate to ensure that the rights of climate-displaced persons are protected. Protection of rights can depend on financial and other resources. Several human rights treaties, including the 1966 International Covenant on Economic, Social, and Cultural Rights (ICESCR), the Convention on the Rights of the Child, and the Convention on the Rights of Persons with Disabilities, identify international cooperation and assistance as necessary to facilitate progressive realization of economic, social, and cultural rights.[79] International support can make providing protection more feasible.

Principle 8b

Principle 8b grants states the "right to seek cooperation and assistance," which can encourage them to ask for the support they need to care for their people. A failure to receive assistance does not relieve states of their duties under the Peninsula Principles. On a practical level, however, assistance can help ensure that states are able to meet their responsibilities, especially if they lack capacity "to adequately prevent and respond to climate displacement."[80]

Principle 8b, which references states' obligations to deal with climate displacement "within their territory," applies only to affected states. Although the Pinheiro Principles do not articulate an identical right, they urge states to request international assistance when necessary.[81] Other international instruments, most notably in the disarmament field, have provisions that more closely resemble that in the Peninsula Principles. For example, the 1997 Mine Ban Treaty and the Convention on Cluster Munitions contain international cooperation and assistance articles that open by granting affected states the right to seek help from other states.[82]

The assistance sought under Principle 8b should be used to implement the Peninsula Principles. The provision says that states should exercise their right in order to fulfill "their obligations to prevent and respond to climate displacement." Prevention and response cover the range of activities the Principles oblige states to take.

Affected states may seek assistance from "other States and relevant international agencies." Developed states are the ones most likely to have resources to provide substantive assistance, although, given the many forms of assistance, any state should be capable of assisting in some way. There are currently no international agencies dedicated specifically to climate displacement, but existing ones could modify their mandates or new ones could be created. For example, the UN High Commissioner for Refugees (UNHCR), the lead agency dealing with the protection and assistance of refugees, could formally extend its programs to encompass climate-displaced persons. Alternatively, the organization could be used as a model for a future agency dedicated to preventing and minimizing the effects of climate displacement.

The UNHCR has already de facto expanded its work to help internally displaced persons, and it has provided support during some natural disaster situations. Its "original mandate does not specifically cover [internally displaced persons], but because of the agency's expertise on displacement, it has for many years been protecting and assisting millions of them."[83] Displacement through climate change and other natural disasters is similarly outside the UNHCR mandate. Nevertheless, the organization has expressed an interest in the issue because environmental degradation can exacerbate other causes of displacement and because climate-displaced persons have "needs and vulnerabilities" comparable to those of traditional refugees and internally displaced persons.[84] According to a senior UNHCR official, as of 2013 the agency was responding on a case-by-case basis to natural disaster situations, which could include those attributable to climate change, but it often assisted if it had the capacity and the affected state requested help.[85] The official said the organization would "continue to respond favourably to requests for involvement in natural disaster scenarios" under certain conditions.[86] The fact that the UNHCR is more likely to provide protection and assistance when the affected state requests it illustrates the importance of Principle 8b.

Principle 8c

This provision lays out the duty of the international community to provide cooperation and assistance. The provision also protects the sovereignty of affected states because it specifies that the international community should give its assistance to those states that have requested it. The UNFCCC recognizes the importance of preserving national sovereignty in the context of international cooperation. Its Preamble "[r]eaffirm[s] the Principle of sovereignty of States in international cooperation to address climate change."[87]

Principle 8c stresses the need to assist affected states that do not have the capacity fully to implement the Peninsula Principles. According to the Inter-Agency Standing Committee, a humanitarian coordination body, "Where the capacity and/or willingness of the authorities to fulfil their responsibilities is/are insufficient, the international community plays an important role in supporting and complementing the efforts of the State."[88] Principle 8c explains that the international community should give assistance "in particular where the requesting State is unable to adequately prevent and respond to climate displacement." The provision helps ensure implementation of all the Principles, and thus promotes the well-being of people facing or experiencing climate displacement. The possibility of outside support also provides an incentive for affected states to adopt the Peninsula Principles.

As discussed earlier, cooperation and assistance should come from both donor states and relevant international agencies. They can act "either separately or together." For example, a state could provide money or other assistance directly to the affected state. Alternatively, a state could contribute financial resources to a fund that an agency draws from in providing assistance.

Principle 8d

While offers of cooperation and assistance from states and international agencies are crucial, affected states, regardless of whether they have requested help, should play their part by accepting it. Principle 8d places this duty specifically on states "unable to adequately prevent and respond to climate displacement" because they are in the greatest need of aid. Affected states' consent is necessary for assistance to reach their people; due to international rules of sovereignty, assisting parties generally cannot infringe on the internal affairs of another state. Precedent for Principle 8d appears in the Convention on Cluster Munitions, which obligates states seeking and receiving assistance to "facilitate timely and effective implementation of this Convention" and to cooperate with donor states.[89]

Conclusion

The General Obligations draw on human rights and international environmental law to address an emerging humanitarian problem. This interdisciplinary approach is appropriate for a novel situation that arises when threats to humans and the environment are intertwined at a global level yet fall through the cracks of existing international law.[90] While there is no single legal instrument dedicated specifically to climate displacement, the General Obligations borrow concepts and language from well-accepted treaties, Principles, and practitioners' guidelines. Reliance on these sources demonstrates consistency with existing legal standards and increases the legitimacy and authority of the General Obligations. Because states have accepted similar provisions in the past, it should be easier for states to adopt them now in the context of climate displacement.

The General Obligations make clear that the first priority of the Peninsula Principles is to preempt climate displacement. Principles 5 and 6 oblige states to take both legal and concrete steps to keep people in their homes. Principle 5 mandates compliance with international law, which can help reduce the need for displacement due to climate change. Principle 6 requires states to minimize forced migration by providing adaptation assistance and protection for human rights.

Principles 7 and 8 concern tools with which to implement the Peninsula Principles. Principle 7 requires affected states to establish a national legislative and structural framework to operationalize the Peninsula Principles. Principle 8 calls on donor states and agencies to provide cooperation and assistance to help affected states meet their responsibilities and on affected states to accept this help.

Together, the General Obligations lay the foundation for implementing the specific Principles that follow. When states operationalize these provisions, they should prioritize preemption while preparing to take any necessary remedial steps. As affected states fulfill their duties, they should utilize and refine the national laws and structures they were required to institute. Concurrently, donor states should bear in mind and adhere to their responsibility to provide support. While valuable in their own right, the General Obligations are also essential to the realization of the Peninsula Principles as a whole.

Notes

1 The Introduction to the Peninsula Principles articulates their scope and purpose, defines terms, prohibits discrimination, and provides rules for interpretation.
2 Although these four Principles use the verb "should" rather than "shall," given their placement under the heading of General Obligations, this chapter will treat them as obligatory for states that choose to adopt the Peninsula Principles.
3 Guiding Principles on Internal Displacement, addendum to the Report of the Representative of the Secretary-General, Francis M. Deng, submitted pursuant to Commission on Human Rights resolution 1997/39, UN Doc. E/CN.4/1998/53/Add.2, February 11, 1998, Introduction, para. 2.
4 See UN Framework Convention on Climate Change (UNFCCC), opened for signature May 9, 1992, S. Treaty Doc. No. 102–38, 1771 United Nations Treaty Series (UNTS) 107, entered into force March 21, 1994.
5 UN High Commissioner for Refugees (UNHCR), *Internally Displaced People: Questions & Answers*, September 2007, www.refworld.org/docid/47a7078e1. html (accessed June 23, 2014), p. 13. The Peninsula Principles "build on and contextualize" the Guiding Principles: Peninsula Principles, Preamble, para. 3.
6 Guiding Principles on Internal Displacement, Principle 5.
7 See Global Protection Cluster, *Handbook for the Protection of Internally Displaced Persons*, June 2010, www.refworld.org/docid/4790cbc02.html (accessed June 23, 2014), p. 483 (describing the Guiding Principles on Internal Displacement as "the accepted standard for humanitarian and human rights actors both nationally and internationally"). The Global Protection Cluster, *inter alia*, leads "standard and policy setting relating to protection in complex and natural disaster humanitarian emergencies, in particular with regard to the protection of internally displaced persons": Global Protection Cluster, "Who We Are," www.globalprotectioncluster.org/en/about-us/who-we-are.html (accessed June 20, 2014).
8 Rio Declaration on Environment and Development, adopted June 14, 1992, UN Doc. A/CONF.151/26 (Vol. 1), August 12, 1992, Principle 15.
9 UNFCCC, art. 3(3).
10 Ibid.
11 Ibid., art. 1(1).
12 Cancún Adaptation Framework, agreed to on December 11, 2010, in Report of the Sixteenth Conference of the Parties to the UNFCCC, Part Two, FCCC/CP/2010/7/Add.1, March 15, 2011, para. 14(f). The Cancún Adaptation Framework consists of paragraphs 11–35 of the Cancún Agreements.
13 Guiding Principles on Internal Displacement, Principle 5.
14 Ibid., Principle 6(1).
15 Ibid., Principle 6(2)(d).
16 Walter Kälin, *Guiding Principles on Internal Displacement: Annotations* (Washington, DC: American Society of International Law and Brookings Institution, 2008), pp. 26–27.
17 Ibid., p. 28. See also International Covenant on Civil and Political Rights (ICCPR), adopted December 16, 1966, G.A. Res. 2200A (XXI), 21 UN GAOR Supp. (No. 16) at 52, UN Doc. A/6316 (1966), 999 UNTS 171, entered into

force March 23, 1976, art. 12(1). In its general comment on Article 12, the Human Rights Committee, the treaty body for the ICCPR, explains that "[e]veryone lawfully within the territory of a State enjoys, within that territory, the right to move freely and to choose his or her places of residence" and that forced internal displacement is prohibited: Human Rights Committee, General Comment No. 27: Article 12 (Freedom of Movement), UN Doc. CCPR/C/21/ Rev.1/Add.9 (1999), paras. 4 and 7.

18 Guiding Principles on Internal Displacement, Principle 6(2)(d) (referring to the only circumstances in which displacement is allowed in disaster situations).

19 See, e.g., Peninsula Principles, Principles 2, 7 9, 10, 11, and 17.

20 Guiding Principles on Internal Displacement, Introduction, para. 2. The UNHCR, the UN's refugee agency, has also used the phrase "individuals, households and communities" in publications. See, e.g., UNHCR, *UNHCR's Role in Support of the Return and Reintegration of Displaced Populations: Policy Framework and Implementation Strategy* (Geneva: UNHCR, 2008), para. 4.

21 Basic Principles and Guidelines on the Right to a Remedy and Reparation for Victims of Gross Violations of International Human Rights Law and Serious Violations of International Humanitarian Law, GA Res. 60/47, December 16, 2005, para. 8.

22 Convention on Cluster Munitions, adopted May 30, 2008, Dublin Diplomatic Conference on Cluster Munitions, CCM/77, entered into force August 1, 2010, art. 2(1).

23 See, e.g., Guiding Principles on Internal Displacement, Introduction, para. 2.

24 The Peninsula Principles define "climate displaced persons" as "individuals, households or communities who face or experience climate displacement": Peninsula Principles, Principle 2c.

25 Global Protection Cluster, *Handbook for the Protection of Internally Displaced Persons*, p. 8. Pastoralists depend on livestock for their livelihood, and because they have to travel to find food and water for their herds, their way of life is "characterized by a high degree of mobility": ibid., p. 511.

26 The Principle also allows for "other measures," thus leaving the door open to additional forms of help not specifically mentioned in the Peninsula Principles.

27 Global Protection Cluster, *Handbook for the Protection of Internally Displaced Persons*, p. 506.

28 Ibid.

29 International human rights law obliges states to

> *respect* human rights, i.e., to refrain from actively violating them; to *protect* such rights, i.e., to intervene and take protective action on behalf of the victim against threats by others or stemming from such a situation; to *fulfil* them, i.e., to provide goods and services necessary to allow people to fully enjoy their rights.
>
> (Inter-Agency Standing Committee, *IASC Operational Guidelines on the Protection of Persons in Situations of Natural Disasters* (Washington, DC: Brookings–Bern Project on Internal Displacement, 2011), p. 5 (hereinafter *IASC Operational Guidelines*))

30 Global Protection Cluster, *Handbook for the Protection of Internally Displaced Persons*, p. 511.

31 Kälin, *Guiding Principles on Internal Displacement: Annotations*, p. 19.

32 The phrase can be read to modify either "provid[ing] adaptation assistance, protection and other measures" or "ensur[ing] that individuals, households and communities can remain in their homes." Regardless, the emphasis of both alternatives on human rights is clear.

33 Guiding Principles on Internal Displacement, Principle 9.

34 UN Declaration on the Rights of Indigenous Peoples, GA Res. 61/295, September 13, 2007, Preamble, para. 7. Only four states (Australia, Canada, New Zealand, and the United States) opposed adoption of this declaration in the UN General Assembly, but they have since reversed their opposition and endorsed it. There were 11 abstentions. See Indigenous Foundations, "UN Declaration on the Rights of Indigenous Peoples," n.d., http://indigenousfoundations.arts. ubc.ca/home/global-indigenous-issues/un-declaration-on-the-rights-of-indigenous-peoples.html (accessed June 5, 2014).

35 UN Declaration on the Rights of Indigenous Peoples, art. 25.

36 Convention concerning Indigenous and Tribal Peoples in Independent Countries, adopted June 27, 1989, entered into force September 5, 1991, art. 13(1) (hereinafter ILO Convention No. 169).

37 Ibid., art. 1(b).

38 Guiding Principles on Internal Displacement, Principle 9.

39 Ibid., Principle 3(1).

40 Nansen Principles, Principle II, in Norwegian Refugee Council/Internal Displacement Monitoring Centre (NRC/IDMC), *The Nansen Conference: Climate Change and Displacement in the 21st Century*, June 7, 2011, www.refworld. org/docid/521485ef4.html (accessed June 24, 2014). The Nansen Principles led to the formation of the Nansen Initiative, which calls for states to develop an international framework for helping people affected by cross-border displacement from climate change. See Walter Kälin, "From the Nansen Principles to the Nansen Initiative," *Forced Migration Review*, no. 41, December 2012, www.refworld.org/docid/50c5c5fb2.html (accessed June 23, 2014), p. 49.

41 UNFCCC, Preamble, para. 9.

42 Kälin, *Guiding Principles on Internal Displacement: Annotations*, p. 19 (citing Article 2(7) of the UN Charter as well as several General Assembly resolutions including Resolution 45/100 of December 14, 1990). The UN Charter, art. 2(7), states, "Nothing contained in the present Charter shall authorize the United Nations to intervene in matters which are essentially within the domestic jurisdiction of any state."

43 Convention on Access to Information, Public Participation in Decision-Making and Access to Justice in Environmental Matters, adopted June 25, 1998, entered into force October 30, 2001, art. 3(1) (hereinafter Aarhus Convention).

44 Convention on the Rights of the Child, adopted November 20, 1989, GA Res. 44/25, annex, 44 UN GAOR Supp. (No. 49) at 167, UN Doc. A/44/49 (1989), entered into force September 2, 1990, art. 4; Convention on the Rights of Persons with Disabilities, adopted December 13, 2006, entered into force May 3, 2008, arts. 4 and 33. The Refugee Convention and its 1967 Protocol do not explicitly require implementation, but they do so implicitly by requiring states parties to report on their implementation laws and regulations: Convention relating to the Status of Refugees (Refugee Convention), adopted July 25, 1951, 189 UNTS 150, entered into force April 22, 1954, art. 36; Protocol Relating to the Status of Refugees, adopted December 16, 1967, 606 U.N.T.S. 267, entered into force October 4, 1967, art. 3. Two major disarmament treaties, the Convention on Cluster Munitions and the Mine Ban Treaty, include similar provisions, which have spawned dozens of national laws that reinforce international obligations: Convention on Cluster Munitions, art. 9; Convention on the Prohibition of the Use, Stockpiling, Production and Transfer of Anti-Personnel Mines and on their Destruction (Mine Ban Treaty), adopted September 18, 1997, entered into force March 1, 1999, art. 9.

45 Scott Leckie, "Welcome to the Peninsula Principles on Climate Displacement within States!," in *The Peninsula Principles on Climate Displacement within States* (Geneva: Displacement Solutions, 2013), p. 10.

46 IASC Operational Guidelines, p. 30.
47 UN Principles on Housing and Property Restitution for Refugees and Displaced Persons, in The Pinheiro Principles: United Nations Principles on Housing and Property Restitution for Refugees and Displaced Persons (Geneva: Centre on Housing Rights and Evictions, 2005), Principle 12 (hereinafter Pinheiro Principles).
48 IASC Operational Guidelines, p. 56.
49 See, e.g., Inter-Agency Standing Committee, IASC Framework on Durable Solutions for Internally Displaced Persons (Washington, DC: Brookings–Bern Project on Internal Displacement, 2010).
50 Vienna Declaration and Program of Action, adopted June 25, 1993, section I, para. 23 (calling for the "achievement of durable solutions" to refugee crises). For UN General Assembly endorsement, see "World Conference on Human Rights," GA Res. 48/121, December 20, 1993.
51 Kälin, Guiding Principles on Internal Displacement: Annotations, p. 126.
52 Ibid. See, e.g., Universal Declaration of Human Rights, adopted December 10, 1948, G.A. Res. 217A(III), UN Doc. A/810 at 71 (1948), art. 13 (stating, (1) "Everyone has the right to freedom of movement and residence within the borders of each state," and (2) "Everyone has the right to leave any country, including his own, and to return to his country."); ICCPR, art. 12(1).
53 Kälin, Guiding Principles on Internal Displacement: Annotations, p. 126.
54 Guiding Principles on Internal Displacement, Principle 28(1). For the right of return for indigenous peoples, see, e.g., UN Declaration on the Rights of Indigenous Peoples, art. 10.
55 Kälin, Guiding Principles on Internal Displacement: Annotations, p. 125. See also IASC Operational Guidelines, p. 56.
56 Guiding Principles on Internal Displacement, Principle 28(1).
57 According to the Handbook for the Protection of Internally Displaced Persons, "[p]eople displaced by natural disasters may no longer be able to return to areas not suitable for habitation": Global Protection Cluster, Handbook for the Protection of Internally Displaced Persons, p. 485.
58 Kälin, Guiding Principles on Internal Displacement: Annotations, p. 129. See Guiding Principles on Internal Displacement, Principle 28.
59 See, e.g., IASC Operational Guidelines, pp. 47 and 56; Pinheiro Principles, Principle 10.
60 Aarhus Convention, Preamble, para. 2, and art. 1. See also Rio Declaration, Principle 10:

> Environmental issues are best handled with the participation of all concerned citizens, at the relevant level. At the national level, each individual shall have appropriate access to information concerning the environment that is held by public authorities … and the opportunity to participate in decision-making processes. States shall facilitate and encourage public awareness and participation by making information widely available. Effective access to judicial and administrative proceedings, including redress and remedy, shall be provided.

61 Aarhus Convention, art. 8.
62 IASC Operational Guidelines, p. 11. See also Pinheiro Principles, Principle 14 (requiring "adequate consultation and participation with the affected persons, groups and communities"); Nansen Principles, Principle X (stating, "The voices of the displaced or those threatened with displacement, loss of home or livelihood, must be heard and taken into account").
63 IASC Operational Guidelines, p. 11.
64 UNFCCC, art. 6(a).

65 Guiding Principles on Internal Displacement, Principle 7(3)(b)–(c).

66 Ibid., Principle 28(2).

67 Aarhus Convention, art. 6(2).

68 See, e.g., UN Declaration on the Rights of Indigenous Peoples, art. 10; ILO Convention No. 169, art. 6; Convention on the Elimination of All Forms of Discrimination against Women, adopted December 18, 1979, GA Res. 34/180, 34 UN GAOR Supp. (No. 46) at 193, UN Doc. A/34/46, entered into force September 3, 1981, Preamble, para 11; Convention on the Rights of Persons with Disabilities, art. 3(c); Convention on the Rights of the Child, art. 9.

69 This paragraph applies explicitly to legislation and does not mention administrative implementation measures. According to Principle 4b, however, the Peninsula Principles should be interpreted "broadly, be guided by their humanitarian purpose, and display fairness, reasonableness, generosity and flexibility in their interpretation." Therefore, the requirement for administrative measures to be consistent with human rights law can be read into Principle 7e.

70 The Preamble to the Peninsula Principles reaffirms the ICCPR, International Covenant on Economic, Social, and Cultural Rights (ICESCR), and Vienna Declaration and Program of Action: Peninsula Principles, Preamble, para. 2.

71 See, e.g., Peninsula Principles, Principles 2d, 3b, 10ei, 11cv, and 11dii.

72 Guiding Principles on Internal Displacement, Principles 8, 18, and 22. The Refugee Convention enumerates even more rights relevant to displacement.

73 For example, the *IASC Operational Guidelines* (p. 9) describe four categories of relevant rights, including:

- rights to life, security and physical integrity, and the protection of family ties during evacuations;
- rights to food, health, shelter, and education;
- rights to housing, land, property, and livelihoods; and
- rights to documentation, freedom of movement, reestablishment of family ties, expression and opinion, and the opportunity to vote.

For another list of rights relevant to persons displaced due to natural disasters, see also Global Protection Cluster, *Handbook for the Protection of Internally Displaced Persons*, pp. 485–486.

74 Refugee Convention, Preamble, para. 4 (referring to the problem of asylum in particular).

75 See, e.g., UNFCCC, art. 4(3). The Peninsula Principles, unlike the UNFCCC, do not classify countries as developing and developed. Either type could be affected by climate displacement, although developing states are generally more vulnerable to the phenomenon and developed states possess more resources to handle it.

76 Nansen Principles, Principle IV.

77 UNFCCC, Preamble, para. 6.

78 For a similar conclusion in the context of the Convention on Cluster Munitions, which requires states parties "in a position to do so" to provide cooperation and assistance, see Bonnie Docherty and Richard Moyes, "Article 6: International Cooperation and Assistance," in Gro Nystuen and Stuart Casey-Maslen, eds., *The Convention on Cluster Munitions: A Commentary* (Oxford: Oxford University Press, 2010), p. 394 ("States may assist in so many ways that they should virtually always be 'in a position to do so.'").

79 See, e.g., ICESCR, adopted December 16, 1966, GA Res. 2200A (XXI), 21 UN GAOR Supp. (No. 16) at 49, UN Doc. A/6316 (1966), 993 UNTS 3, entered into force January 3, 1976, art. 2(1); Convention on the Rights of the Child, Preamble, para. 11 and art. 4; Convention on the Rights of Persons with Disabilities, art. 4(2).

80 Principle 8c and d. As discussed below, Principle 8c urges donor states and agencies to make particular efforts to help affected states with limited capacities.

81 Pinheiro Principles, Principle 104: "States should, when necessary, request from other States or international organizations the financial and/or technical assistance required to facilitate the effective voluntary return, in safety and dignity, of refugees and displaced persons."

82 Mine Ban Treaty, art. 6(1); Convention on Cluster Munitions, art. 6(1). These two treaties also grant affected states the right to receive assistance.

83 UNHCR, "Internally Displaced People: On the Run in Their Own Land," www.unhcr.org/pages/49c3646c146.html (accessed June 24, 2014).

84 José Riera, senior advisor to the director of international protection, UNHCR headquarters, "Challenges Relating to Climate Change Induced Displacement," remarks at international conference entitled "Millions of People without Protection: Climate Change Induced Displacement in Developing Countries," Berlin, January 29, 2013, www.refworld.org/docid/510a3a372.html (accessed June 24, 2014), p. 4. See also generally UNHCR, *Climate Change, Natural Disasters and Human Displacement: A UNHCR Perspective*, August 14, 2009, www.refworld.org/docid/4a8e4f8b2.html (accessed June 24, 2014).

85 Riera, "Challenges Relating to Climate Change Induced Displacement," p. 4. Assistance also depended on whether the UNHCR had "an established presence, programme and relief items in the country": ibid.

86 Those conditions included "an invitation by the disaster-affected country and the Emergency Relief Coordinator" as well as a presence and capacity in the country and "the absence of another agency in country with capacity to take the lead": ibid.

87 UNFCCC, Preamble, para. 9.

88 *IASC Operational Guidelines*, p. 6.

89 Convention on Cluster Munitions, art. 6(10) and (12). See also Mine Ban Treaty, art. 6(6).

90 Bonnie Docherty, "Climate Change Migration and Social Innovation," *Harvard College Review of Environment and Society*, Spring 2014, www.hcs.harvard.edu/~res/wp-content/uploads/2014/02/HCRECS_Spring_14_Final.pdf (accessed October 11, 2014), pp. 22–24.

6 Climate displacement

Preparation and planning

Robin Bronen

This chapter focuses on one of the critical protection gaps addressed by the Peninsula Principles – the design of an institutional framework to prepare and plan for climate-induced relocations and determine when and how they need to occur. Although the Guiding Principles on Internal Displacement and the Inter-Agency Standing Committee (IASC) Operational Guidelines on the Protection of Persons in Situations of Natural Disasters outline the human rights protections for populations displaced by natural disasters, these guidelines are focused on emergency responses and do not address the significant planning needs, and the human rights implications, for relocating populations prior to being displaced.

Peninsula Principles 9, 10, 11 and 13 outline the components of this framework. Incorporating human rights principles identified in the International Covenant on Economic, Social and Cultural Rights (ICESCR) and the International Covenant on Civil and Political Rights (ICCPR), the Peninsula Principles articulate a rights-based approach to guide individuals, households, communities and governments on the steps that must be taken to prepare for a planned relocation. Included in this institutional design, the Peninsula Principles also outline the criteria for informed consent by and participation of climate-displaced populations, one of the most important components of planning and preparation for relocation. Fundamental to the Peninsula Principles' articulation of consent and participation is the right of climate-impacted populations to decide whether, when and how to relocate, including the collective right of communities to make these decisions.

The Principles affirm the importance of protection in place and the right of people 'to remain in their homes and retain connections to the land on which they live for as long as possible'. Principle 13 also articulates the importance of government technical assistance to prevent, prepare for and respond to climate displacement so that affected populations have access to the critical information and resources needed to respond to the climate-induced environmental threats. This exchange of information and resources is particularly important for populations that typically have the least access to these assets, such as indigenous peoples, women, the elderly, ethnic minorities, and those living in poverty (Principle 13biii).

Relocation is complicated, can be severely harmful to people, is a long-term development process and can be expensive to plan and implement. In order to identify and mitigate adverse social impacts and impoverishment risks, planning must begin as soon as it is established that relocation must occur. New multi-level and multi-disciplinary relationships between national, regional, tribal and local government actors must be created in order for them to work in concert together and with climate-affected populations. For these reasons, relocation should only occur as a last resort, if there are no other adaptation strategies to protect people from the climate-induced environmental changes in the places where they reside.

However, in the context of a rapidly changing environment, the ongoing effort to protect people in place may recreate or increase existing vulnerabilities, and preclude longer-term planning and policy changes for enhancing adaptation to climate change (Grannis, 2011; Lewis, 2012). Currently, significant limitations prevent governments from responding effectively and dynamically to climate-induced environmental changes. Governance tools are primarily dedicated to protect people and infrastructure *in situ*, even though this may offer only short-term protection because of recurrent drought and increased rates of sea level rise, and the consequent increased intensity of storm surges, flooding and erosion.

Principles 9 and 13 emphasize the importance for States to design and implement a relocation institutional framework. In order for governance institutions to respond dynamically to the humanitarian needs of populations faced with changing environmental conditions, this framework must be adaptive and enable institutions to prepare for a continuum of responses which includes post-disaster recovery, protection in place (consisting of seawall and shoreline protection), hazard mitigation, and relocation (Bronen and Chapin, 2013; Bronen, 2011).

Principle 9 also emphasizes the importance of developing this institutional framework with the participation of climate-affected populations in order to ensure that the institutional design addresses their concerns. By including climate-affected populations in the design of the institutional framework, the negative impacts of relocation, including the weakening of social networks and traditional kinship ties and the disruption of subsistence and economic systems, may be minimized and potentially avoided. In addition, climate-affected populations can assist governments to develop and incorporate opportunities for sustainable development and improvement of incomes and living standards into the relocation process.

Design of an adaptive governance framework to prepare and plan for climate-induced relocation

The Peninsula Principles outline for States the essential components of an adaptive governance framework to prepare and plan for climate-induced relocations. Three main organizational instruments need to be designed to create this relocation adaptive institutional framework: a relocation policy framework, a relocation organizational framework, and a relocation plan for each individual community.

Relocation policy framework

Principles 9 and 10 articulate the critical elements of a relocation policy framework, which includes the following: the design and implementation of multi-sector and multi-level monitoring and assessment of climate-induced environmental change; the development of social-environmental indicators to determine when relocations need to occur to protect people; the identification and acquisition of relocation sites; the organizational structure, roles and responsibilities for governmental and non-governmental agencies when facilitating a relocation process; and the financial mechanisms to fund the relocation effort. Peninsula Principle 9e encourages States to integrate the laws, rights and procedures designed for this relocation policy framework into already existing legal and institutional systems and consult and collaborate with regional and local authorities in the development of this institutional framework.

Human rights Principles

The Peninsula Principles affirm the importance of incorporating already recognized human rights Principles into the relocation policy framework (Principle 10e). These Principles should include the right to relocate when climate-induced environmental change threatens the lives of community residents and erosion control and flood relief cannot provide protection; the right to life, which mandates a nation state government to protect its citizenry from climate-induced environmental threats (UN, 1948; UN Human Rights Committee, 1982); and the right to self-determination[1] to empower communities during the relocation process and ensure that the relocation is community-based and community-guided (Risse, 2009). In order to further this last Principle, affected communities must be designated as key leaders in the relocation process.

Principle 10 also affirms the importance of protecting the right to adequate housing (UN Committee on Economic, Social and Cultural Rights, 1992), potable water, improved standard of living, subsistence rights and access to customary communal resources defined in the UN International Covenant on Economic, Social, and Cultural Rights and the UN Declaration on the Rights of Indigenous Peoples. These rights must be protected during displacement as well as at the relocation site. The next section begins with a discussion of informed consent and participation, outlined in Principle 10, because of its importance in framing the relocation policy framework. Next, the section describes two critical institutional components of the relocation policy framework, social-ecological indicators and multi-level environmental and sociological monitoring and assessment. These two institutional components form the foundation upon which the right to self-determination and informed consent and participation are incorporated into the relocation planning process. The section concludes with a discussion of the economic, social and cultural human rights Principles which must be embedded in the relocation policy framework.

Principle 10: Informed consent and participation

Principle 10 articulates for States a framework to incorporate participation and consent and the right to self-determination into the relocation planning process and prioritizes a State's responsibility to relocate those communities, individuals and households choosing to relocate. By prioritizing the relocation of those who choose to relocate and stating that no relocation shall take place unless both displaced populations and host communities provide full and informed consent for such relocation, Principle 10 embeds the notion of voluntariness and the right to self-determination into the relocation decision-making process. Principle 10 balances the tension between the right to self-determination and consent with a government's responsibility to protect people through preventive relocations by stating that relocation can occur 'without such consent in exceptional circumstances when necessary to protect public health and safety or when individuals, households and communities face imminent loss of life or limb'.

The Peninsula Principles thus divide the issue of who makes the decision to relocate into three dimensions: individuals, households and communities who decide to relocate; government actors who decide relocation must occur because populations are imminently threatened by climate-induced environmental change; and government actors who decide relocation must occur when those threats are not imminent. It is in this last context that the right to participate and consent in the decision to relocate is paramount.

Participation and informed consent are vague terms to describe the type of communication and decision-making which needs to occur between climate-affected populations and government agencies to ensure that climate-affected populations have the critical information needed to demonstrate that relocation is warranted. Principle 10ei states that affected individuals, households and communities, both those displaced and those hosting displaced persons, must be fully informed about the relocation so that they can actively participate and implement relevant decisions, laws, policies and programmes designed to ensure respect for and protection of housing, land and property and livelihood rights.

In order for informed consent to be meaningful, institutional arrangements need to be designed and implemented so that the participation of climate-affected populations can commence at the earliest stages of relocation planning and continue throughout the relocation process. The relocation policy framework must describe the methods through which climate-affected populations can participate and make decisions throughout relocation planning and implementation, and the mechanisms through which they can communicate their concerns to government actors. These institutional mechanisms need to be culturally and linguistically appropriate and ensure that vulnerable populations such as Indigenous peoples, ethnic minorities and women are adequately included in the communication and participation strategy to achieve informed consent. The relocation policy framework should also describe the methods for disseminating information about all stages of the relocation process, ensure that the document is disseminated in a form, manner and language which are understandable to the community being relocated and also address issues about the quality of the process, such as leadership and representation (World Bank, 2004).

Principle 10e includes host populations in the participation and informed consent process because of the importance of protecting their human rights, avoiding conflict between relocated populations and host communities and ensuring that host populations do not experience a fall in living standards and economic opportunities.

While informed consent and participation are fundamental components of a government-initiated relocation process, they do not constitute and are not equivalent to self-determination and may not be sufficient to protect the human rights of those threatened by climate-induced environmental change. Self-determination ensures that those affected by the relocation and who best know their economic, social, cultural and physical surroundings make decisions to avoid the social fragmentation and impoverishment associated with involuntary relocations. For this reason, the Peninsula Principles prioritize a State's responsibility to relocate communities which *decide* to relocate, as described below.

Self-determination

In the context of climate-induced environmental change that threatens the lives, livelihoods and habitability of communities, self-determination means that individuals, households and communities have the right to make fundamental decisions about when, how and if relocation occurs. The Peninsula Principles affirm this already recognized right. The failure to fully consider the welfare of climate-affected populations and empower people of a community to make decisions regarding critical elements of a relocation, including site selection and community lay-out, are the principal reasons why relocations have been unsuccessful (Abhas, 2010). The relocation process must ensure that socio-cultural institutions remain intact and that families, communities and ethnic groups, including Indigenous populations, remain together during the relocation process if they make this decision. If they are not able to remain together, they must decide who should relocate together and how the relocation occurs. For these reasons, individuals, households and communities faced with climate-induced ecological threats must have the authority to decide whether or not to relocate. Collective self-determination ensures that communities are empowered to make the critical decisions affecting their relocation.

The concept of self-determination has evolved since the creation of the United Nations in 1945 when the Principle initially was interpreted to apply to the right of independence, non-interference and democracy of a nation state in relation to other nation state governments (Broderstad and Dahl, 2004; UN Committee on Economic, Social and Cultural Rights, Fact Sheet, 2006). Both the International Covenant on Economic, Social and Cultural Rights and the International Covenant on Civil and Political Rights establish that 'all peoples have the right to self-determination' by virtue of which 'they freely determine their political status and freely pursue their economic, social and cultural development'. The inclusion of the right to self-determination in both treaties indicates that its importance spans all political, civil, economic, social and cultural rights.

More recently the concept of self-determination has included self-government institutions in Indigenous communities (Broderstad and Dahl, 2004; UN Committee on Economic, Social and Cultural Rights, 2006). The UN Declaration on the Rights of Indigenous Peoples affirms that Indigenous peoples possess collective rights, indispensable for their existence and well-being, including the right to collective self-determination and the collective right to the lands, territories and natural resources they have traditionally occupied and used (UN, 2007: art.1). The collective right to self-determination ensures that Indigenous communities can determine their own identity, belong to 'an indigenous community or nation, in accordance with the traditions and customs of the community or nation concerned' and make decisions about internal and local affairs (UN, 2007: arts 9, 33). The Declaration also provides that Indigenous peoples have the right to freely define and pursue their economic, social and cultural development. Similarly, the Convention for the Safeguarding of the Intangible Cultural Heritage also affirms the collective rights of communities to safeguard and respect their cultural heritage (UN, 2006: art. 1). For Indigenous communities, kinship relationships are essential to cultural identity.

While these international documents articulate the concept of collective human rights to Indigenous identity, this concept can be extended to other distinct communities where people can demonstrate kinship ties, social network connections and the desire to remain together at the relocation site. By articulating the collective right to self-determination to choose to relocate, the Peninsula Principles affirm the importance for climate-affected populations of relocating as a group, if they choose, to preserve community cohesion and to avoid the break-up of social networks and kinship ties, one of the most pernicious negative outcomes of development-forced displacement (Smith 2009).

Principle 9: Risk monitoring and assessment

In order to operationalize this manifestation of the right to self-determination, Principle 9 describes the design and implementation of a multi-level risk monitoring and assessment mechanism. This mechanism means that climate-affected populations assess and document local environmental changes and sociological impacts and vulnerabilities in collaboration with regional and national government actors. By including households in the data collection process, Principle 9 ensures that those most directly affected by environmental change actively participate in the gathering of data during the risk assessment process (May and Plummer, 2011).

This local-level monitoring and assessment can provide a long-term historical perspective and an understanding of the connections between people and the environment, and provide the mechanism to foster community empowerment, promote human rights protections and encourage transparent decision-making processes – all components of good governance (Alfredsson, 2013). Climate-affected populations can also use this monitoring and assessment to determine whether climate change risks can be mitigated at the original community location or whether relocation needs to occur.

Government agencies can then enhance this local data collection by modelling likely climate displacement scenarios (including timeframes and financial implications) and identifying characteristics of possible relocation sites for climate-displaced persons (Kannen and Forbes, 2011). In this way, both residents and government agencies together can engage in a collaborative process of understanding climate change impacts on community habitability, build local adaptive capacity and implement a dynamic and locally informed institutional response (Kofinas, 2009; Armitage and Plummer, 2010; Berkes, 2009).

Collaborative governance involves cross-boundary collaboration between different levels and sectors of governmental and non-governmental organizations to resolve or address a societal problem that cannot be accomplished without this interdependence (Emerson *et al.*, 2011). Principled engagement is one of the critical elements of collaborative governance and involves a dynamic social learning process. This process can also help reduce the tension between the right of people to remain and the duty of governments to protect life, which may require the relocation of people against their will (Ferris, 2012).

Several issues demonstrate the importance of multi-level risk monitoring and assessment. First, there is no standardized mechanism or criterion to determine whether and when populations need to be relocated due to environmental change. Understanding the rates of climate-induced environmental change is critical in order for individuals, communities and governments to adapt and determine when protection in place is no longer possible and community relocation is required (Kofinas, 2009; May and Plummer, 2011; Winkler *et al.*, 2013). These assessments can thus be the tool to determine whether and when relocation needs to occur to protect the health and well-being of climate-affected populations. Social ecological indicators, discussed below, can be used to assess vulnerability and guide the design of adaptation strategies for communities and government agencies in order to transition from protection in place to community relocation.

In addition, the assessment of when a climate-induced relocation needs to occur requires a dynamic risk monitoring and assessment process closely connected with changes in the environment that affect the health and well-being of community residents. Consistent monitoring of environmental changes and the impact of these changes on individuals, households and the larger community offers the opportunity to capture the dynamic nature of community vulnerability and resilience. Exposure and vulnerability to climate change vary across temporal and spatial scales, and depend on economic, social, geographic, demographic, cultural, institutional, governance and environmental factors (IPCC, 2012). By monitoring slow on-going environmental processes and environmental change that occurs in the aftermath of an extreme weather event, individuals and communities will be better able to understand the progressive nature of climate change. This in turn will perhaps allow them to seek adaptation strategies other than return and protection in place.

Finally, adaptation responses require information about the local environment. Global, regional and national climate change assessments have generally aggregated information above the level of resolution required for effective community policy. Local landscape change can influence microclimate conditions and outweigh the influence of larger geospatial analyses of long-term climate change predictions. For example, at local and regional levels, sea level rise varies and may exceed averaged global projections, depending on a variety of reasons, including the topography and geological factors causing land to subside in certain regions (Sallenger *et al.*, 2012). Consequently, it is critical that decision makers at the local level understand how their particular locality will be affected by global and regional projections of climate-induced environmental change and have the governance tools to effectively identify and evaluate the best policy options to adapt to their local context. Similarly, it is also important for government agencies that may not have access to local information to understand this local scenario and integrate this information into regional or national models of climate change scenarios (Lewis, 2012).

Components of a social-ecological assessment and monitoring tool

The social-ecological assessment and monitoring tool can include environmental monitoring of erosion rates and sea level rise as well as frequency of extreme weather events. The monitoring of erosion may be a critical component of this assessment tool as it is a primary threat to coastal and riverine communities (Woodward-Clyde Consultants, 1998). Monitoring health impacts of climate-induced environmental change is also critical. Similar to the monitoring of environmental change, preventing negative health outcomes requires a local-scale understanding of the type, timing and rate of change, as well as the direct and indirect health effects (Brubaker *et al.*, 2011). Integrated health assessments can systematically identify and quantify the many pathways through which climate change can affect health in different social and ecological contexts. The World Health Organization suggests that a natural point of entry for health impact assessments is during the planning process for climate-induced relocations because the assessment can put key health issues in front of the policy-makers who directly influence the implementation of plans (Winkler, 2013).

Social-ecological indicators

Principle 9f outlines an essential component of this relocation adaptive governance framework: the identification of social-ecological 'indicators that will, with as much precision as possible, classify where, at what point in time, and for whom, relocation will be required as a means of providing durable solutions to those affected'. Governmental and non-governmental actors must know when to collectively and collaboratively shift from the traditional, 'protect in place' post-disaster recovery response to a community relocation process.

These indicators need to be specific to ecosystems; geographical regions; and social, political and economic systems. To determine which communities are most likely to require relocation, a complex assessment of the vulnerability of a community's ecosystem to climate change, as well as the stability of its social, economic and political structures, must be considered. For example, the indicators of socio-ecological vulnerability demonstrating that relocation is required could include:

1   repeated loss of community infrastructure;
2   imminent danger to the community from ongoing ecological changes and repeated random extreme weather events;
3   no ability for community expansion;
4   numbers of evacuation incidents and numbers of people evacuated;
5   predicted rates of environmental change (e.g. sea level rise) modeling of regional and global climate change scenarios;

6   repeated failure of hazard mitigation measures;
7   a lack of viable access to transportation, potable water, communication systems, power, and waste disposal; and
8   decline in socio-economic indicators, including food security, loss of livelihood, and public health (Bronen, 2011).

These indicators can be used to determine whether and when relocation is required to protect the safety of community residents from climate-induced environmental change. In this way multi-level government institutions can work in concert to create this shift from protection in place to relocation, allocate governmental technical assistance and funding, and encourage the efficient use of resources for long-term protection (Bronen, 2011).

Social, economic and cultural human rights

The relocation policy framework must ensure that social, economic and cultural rights are protected.

Right to relocate

The right to relocate is a fundamental component of the right to self-determination. The right to relocate, like other human rights, is an entitlement when relocation is the only feasible solution to protect the human right to life as well as the right to basic necessities inherent in living a life of dignity (Moyn, 2010).

Several international legal instruments support the right to relocate, including the Universal Declaration of Human Rights and the Pinheiro Principles, which both state that everyone has the right to freedom of movement and residence (COHRE, 2007: Principle 9.1). Although this right has been interpreted to mean that no one shall be arbitrarily or unlawfully forced to remain within or leave a territory, this right also includes the right to movement when threatened by environmental events (COHRE, 2007: Principle 9.1). In addition, the human rights guidelines to respond to natural disasters interpret the right to life to mean that people affected by natural disasters should be allowed to relocate to other parts of the country (Brookings-Bern Project on Internal Displacement, 2011).

Right to basic necessities

Principle 10e incorporates the human right to basic necessities essential for a life of dignity by affirming that States should make certain that

> basic services, adequate and affordable housing, education and access to livelihoods (without discrimination) will be available for climate

displaced persons in the host community at a standard ensuring equity between the host and relocating communities, and consistent with the basic human rights of each.

Relocation should not diminish the living standards of the affected communities. Articles 14 and 23 of the UN Declaration on Indigenous Rights affirm collective rights to the fundamental freedoms articulated in international human rights law and specifically include the 'right to be actively involved in developing and determining health, housing and other economic and social programs affecting them and to administer such programs through their own institutions'. These fundamental freedoms include the collective right to basic necessities, which, at a minimum, means that relocated communities must have access to (1) food and water; (2) housing; and (3) adequate health. In addition, relocation sites can enhance living standards by improving housing and access to potable water and electricity.

Right to subsistence and food

Human rights doctrine explicitly states that the right to food and the right to be free from hunger are indispensable to human dignity and critically connected to other fundamental rights (UN Committee on Economic, Social and Cultural Rights, 1999).

States have the primary responsibility to promote and protect the right to food throughout the relocation process (UN Human Rights Committee, 2011). This obligation has three components (UN Committee on Economic, Social and Cultural Rights, 1999: paragraph 15). States are prohibited from taking any actions that prevent individuals' access to food. States also have the obligation to 'strengthen people's access to and utilization of resources and means to ensure their ... food security' (UN Committee on Economic, Social and Cultural Rights, 1999: paragraph 15). Finally, States have the duty to provide food, '[w]henever an individual or group is unable, for reasons beyond their control, to enjoy the right to adequate food by the means at their disposal'. This obligation specifically applies to victims of natural disasters (UN Committee on Economic, Social and Cultural Rights, 1999: paragraph 12).

The right to subsistence, an element of the right to self-determination defined in both the ICCPR and the ICESCR, is one of the essential human rights connected to the right to food (International Covenant on Economic, Social and Cultural Rights, 1966). For many communities, subsistence agriculture, hunting and fishing are the primary means of access to food. Climate-affected populations must be able to determine how and whether they will be able to continue these activities in the relocation process if climate-induced environmental change, including sea level rise which increases flooding and saline intrusion, and changes in precipitation, threatens their ability to engage in subsistence food harvesting (Holthus *et al.*, 1992; Boege, 2011).

For Indigenous peoples, the right to food is a collective right, and fundamentally connected to sovereignty, rights to land and territories, health, subsistence, treaties, economic development and culture (Office of the United Nations High Commissioner on Human Rights, 2008). For these reasons, it is critically important for them to choose their relocation site.

Right to work/economic development/improved standard of living

Principle 10d affirms that States should 'adopt measures that promote livelihoods, acquisition of new skills and economic prosperity for both displaced and host individuals, households and communities'. The relocation policy framework must also create the opportunity to improve livelihoods and standards of living while implementing sustainable development strategies as part of the relocation process. Human development goals which improve the economic and social conditions of community residents, including in the areas of education, employment, vocational training and retraining, housing, sanitation, health and social security, must be incorporated into community relocation planning (UN, 2007: Article 21). Relocation can offer opportunities to improve living standards, rather than merely recreating poverty in new surroundings. Relocation must be seen not simply as moving people to new sites but as a long-term development process of rebuilding livelihoods of displaced people. People should be moved where possibilities exist or could be created to rebuild livelihoods that are lost due to resettlement.

Economic development opportunities – non-land-based options built around opportunities for employment or self-employment – should be provided in addition to cash compensation for land and other assets lost. The burden of poverty falls disproportionately on women, so it is essential to increase their income-earning opportunities, their food security and their access to social services during the relocation process. If poverty reduction is to be sustainable, institution-building and investing in local capacity to assess poverty and to analyse, design, implement and finance programmes and projects are essential (World Bank, 2004). By incorporating these human rights, the relocation process can enhance the resilience capacity of climate-impacted populations by addressing socio-economic issues, such as lack of economic development, which contributes to the vulnerability of communities.

Right to water

Relocated communities must have sufficient amounts of water for their basic household needs, including drinking, cooking, producing food and hygiene. The human right to water is essential for leading a life of human dignity and is indispensable to the realization of all the human rights related to basic necessities, and fundamental for life and health (Office of the United Nations High Commissioner on Human Rights, 2007).

The right to water is not explicitly mentioned in the International Covenant on Economic, Social and Cultural Rights, but is interpreted to be implicit in the right to an adequate standard of living and health (UN Committee on Economic, Social and Cultural Rights, 2002). The Committee on Economic, Social and Cultural Rights (2002: Articles 11 and 12) defines the right to water as the equal and non-discriminatory right of everyone to access sufficient, safe and affordable water for personal and domestic uses. Governments are prohibited from interfering in access to safe drinking water and are obligated to implement strategies to ensure that there is access to water (Office of the High Commissioner of Human Rights, 2007).

Non-discrimination

Relocation must not be conducted in a discriminatory manner of any kind. All articles of non-discrimination articulated in the Guiding Principles on Internal Displacement are incorporated into the Peninsula Principles.

Host populations

The human rights of host communities must also be protected to ensure that they benefit from any relocation, preserve or improve their standard of living, and to prevent conflicts and competition with the displaced populations. Host communities receiving displaced persons should be provided with timely and relevant information about the relocation so that they can determine whether the relocation will occur in their location. States should offer host community residents opportunities to participate in planning, implementing and monitoring the relocation. Grievance procedures need to be established for the host community. Host populations may experience shortages of water, sanitation, shelter and essential health services as a result of the increase in population. Schools may also be overburdened if there is an influx of displaced students. Infrastructure in host communities should be improved, restored or maintained at levels of service similar to those provided to the relocated population. Similarly, if existing infrastructure or services in the host community are of a lower standard than those provided to the relocated population, the host community's infrastructure should be upgraded.

Embedding these human rights Principles in the relocation policy framework will ensure that climate-affected populations and host communities are empowered to make decisions fundamental to the relocation process and that the human rights essential to live a life of dignity are protected.

Principle 11: Land identification, use and habitability

The relocation site is the critical issue for all peoples needing to relocate. The relocation policy framework must address this issue. Principle 11 highlights the key issues that must be considered when identifying the places where populations will relocate. Land-centred relocations for housing, agriculture and subsistence purposes can be complicated. In densely populated areas, land may not be available for relocation, not habitable or not sufficient for subsistence activities. In addition, in places where land is available and could provide a direct land replacement, the relocation may encourage conversion of forest to agricultural land. The complexity of this issue requires States to 'identify, acquire and reserve sufficient, suitable, habitable and appropriate public and other land to provide viable and affordable land-based solutions to climate displacement, including through a National Climate Land Bank' (Principle 11ai).

Principle 11 incorporates three human rights Principles which apply to the right to housing in a relocation process: (1) the right to replacement housing; (2) the right to habitable housing; and (3) the right to choose the place of one's residence. The right to habitable housing means the housing provides adequate protection from environmental hazards, including weather, and is located away from hazardous zones. The human right to property, defined in Article 17 of the Universal Declaration of Human Rights, includes the right to land ownership. Governments need to enact policies and laws to ensure that housing, land and property restitution procedures are embedded within a legally sound, coherent and practical framework. These procedures should bring displacement to a permanent, sustainable and just end and be compatible with international human rights, refugee and humanitarian law and related standards (COHRE, 2007: 11.1). Principles which must be included within these laws include non-discrimination in housing restitution (COHRE, 2007: 4.1–4.2). All displaced persons must also have the right to full and effective compensation as an integral component of the relocation process to ensure that all aspects of the loss of housing and land are rectified and that the relocation does not cause further impoverishment (Leckie, 2009).

Land-centred remedies for loss of land-based incomes should be a priority. Principle 11 articulates the need to create geographically relevant standardized criteria to evaluate the habitability and feasibility of the relocation site. These criteria should include:

1   current land use, including for subsistence;
2   restrictions associated with the land, such as environmental protections;
3   habitability of the land, including accessibility, availability of water, climate change vulnerabilities (e.g. vulnerability to storm surges or thaw of ice-rich permafrost);

4   feasibility of subsistence/agricultural, hunting and fishing use, including soil surveys to determine the capacity of the soil to support agriculture and agro-meteorological surveys of rainfall, temperature and groundwater;

5   a detailed demographic and land-ownership survey of the host community (Bronen, 2011).

Specifically defining these criteria is essential so that the community being relocated and the government agencies providing technical assistance are in agreement in regard to the habitability of the relocation site. Any disagreement over the relocation site will only serve to delay and impede relocation efforts.

Land ownership is critical in the relocation process. Relocation of an entire village to a new location creates complex and unique public and private property rights issues that need to be addressed in the relocation planning process. Local governments will need to determine land tenure issues, such as whether property will be common, public or privately held, and land title allocation between prospective community residents, businesses and government entities.

Principle 11 also recognizes that States need to include institutional mechanisms to ensure that housing, land, property and livelihood rights will be met for climate-displaced persons

> who have informal land rights, customary land rights, occupancy rights or rights of customary usage, and assurances that such rights are ongoing; and assurances that rights to access traditional lands and waters (for example, for hunting, grazing, fishing and religious purposes) are maintained or similarly replicated.
>
> (Principle 11d)

The right to land-ownership restitution requires that specific arrangements be made to recognize claims to land title and ownership, especially for Indigenous peoples who may not have formal land titles and who may own land collectively (Brookings-Bern, 2011).

The relocation of communities also requires many types of government approvals and permits due to the potential construction of multiple major facilities, including airports, barge landings, schools, health clinics and housing. No one government agency is responsible for the construction of all these facilities. The process framework needs to identify the permitting requirements for relocation and develop a plan to fulfil these legal obligations. In addition, community usage of the old site, which may provide critical access to subsistence resources, cemetries or historical sites, needs to be clarified.

Relocation operational framework

The relocation policy framework will outline the basic elements of a relocation operational framework to identify:

1   whether a specific community must relocate or whether adaptation strategies can be used to avoid relocation;
2   the steps governmental and non-governmental agencies must take to implement a relocation process, such as a community socio-ecological assessment documenting that relocation is warranted, a community-wide vote or survey demonstrating community commitment to relocate, and a relocation site selection process which includes community approval of the site chosen;
3   the organizational arrangements between multi-disciplinary governmental and non-governmental agencies;
4   the funding mechanisms for relocation.

Under this new framework, lead relocation agencies would be responsible for implementing two essential organizational components to address the unique issues that arise each time a community relocates: an operational framework for relocation planning and implementation, and an operational framework for the actual relocation. These lead relocation agencies will provide overall authority to guide multi-disciplinary and multi-level governmental and non-governmental teams of agencies involved in community-specific relocation plans. Legislation should specifically outline the institutional framework and funding for the relocation process.

During the planning and implementation phase it will be critical to examine all people/communities that will be impacted by relocation (not only those who are actually being displaced), and determine whether relocation of climate-affected populations will deprive the original community of the 'critical mass' needed to sustain economic productivity. With the relocation of some members of the community, those remaining in the community may find it impossible to recreate their livelihoods to the same level as before the relocation.

Operational relocation framework to guide community-specific relocations

The operational relocation framework should:

1   outline the comprehensive master relocation plan;
2   identify the staffing patterns required for relocation;
3   develop a capacity-building plan for the relocation staff (if necessary);
4   develop coordination arrangements among relevant agencies and determine the government agencies to be involved in the relocation;

5 monitor the health and well-being of community residents during the relocation process;
6 design and implement the process for gathering and disseminating information;
7 create an overall timeframe for completing the relocation and decommissioning the old village site, if necessary (Bronen, 2011).

Principle 10 outlines the components of a master relocation plan, which should:

1 identify key stakeholders involved in the community relocation to prepare and implement specific components of the relocation;
2 outline the mechanisms for stakeholder coordination, which include government agencies, people who will be relocating and those who will be affected by the relocation, and civil society groups with an interest in the relocation;
3 define the role of the existing community's governance institutions in the relocation process;
4 develop a land acquisition process and a process to provide transitional and permanent housing;
5 describe the responsibilities and procedures for making relocation decisions;
6 identify regulatory and permitting requirements and determine how each will be met;
7 identify the mechanisms for making modifications to the relocation strategic plan during implementation;
8 select criteria to determine eligibility for relocation assistance;
9 identify measures to improve or restore livelihoods and living standards;
10 describe methods to resolve potential conflicts or grievances, including judicial and traditional dispute settling mechanisms;
11 establish monitoring arrangements;
12 identify infrastructure needing replacement;
13 preserve existing social and cultural institutions and places of relocated populations.
14 Establish a time frame to decommission the original location to ensure that infrastructure is maintained until populations are relocated.

Role of existing local governance institutions

Planning challenges can arise because of the lack of clear statutory guidance about the role of local government in the relocation process. First, the existing community's government may have no authority to make decisions at the relocation site. Second, it may be necessary to define and structure the relationship between the owner of the relocation site and the future government of the new community. Without clearly defining the governance authority at the relocation site, decision-making at the local level may delay the relocation process – or, in the most extreme cases, make it impossible for the local government to have any authority to make decisions connected with the relocation site. Similarly, when a village selects a relocation site that it owns, but access to the site requires moving through property owned by other entities, there must be a process to define the relationship and identify a governing authority responsible for negotiating transit rights (Bronen, 2011).

In order to resolve these issues, the existing community's government must have the authority to be a key leader and decision-maker in the relocation process. The community-specific relocation plan needs to identify the steps that a local government must take to continue in its governance role during the relocation process. The authority to govern may be based on the connection to a defined population or to a defined territory. Clear statutory guidance needs to outline the mechanism that the governing authority of the existing community will use to continue in its governance role over the relocation site.

Capacity building for relocation staff

Relocation places enormous burdens on governance structures. State and local governments are typically structured and staffed to deal with the business of governing established and existing communities. Relocation involves a lot more work than overseeing an existing community. An operational relocation framework needs to address relocation staffing issues so that local government institutions have sufficient resources to plan and implement the relocation.

Principle 13 emphasizes the importance of capacity-building initiatives to build the relocation capacity of the government-implementing agency and communities. Government agencies may lack the technical, organizational or financial capacity to implement relocation plans and may also lack the necessary legal authority or political commitment. Coordination between agencies and the establishment of clear lines of responsibilities are also essential. Funding also needs to be designated to hire and train staff at all levels of government involved in the relocation process.

Assessment of vulnerabilities and impacts of relocated populations

A social and economic census of the climate-affected population to be relocated is critical, even when relocation is chosen by an affected population. Relocation places enormous stress on community residents. Baseline data that document the health and socio-economic status of community residents are critical to the relocation process in order to monitor the health and well-being of community residents. In addition, the relocation process can incorporate special provisions to ensure that the needs of all residents, including the elderly, children and those with medical conditions, are addressed.

Socio-economic surveys can identify and explain the demographics of the people who will be affected, along with the division of labour in households and the community, social networks, kinship connections and local institutions, education and health information, a preliminary valuation of the property, an estimate and source of their annual income and access to common resources. These surveys can also identify the infrastructure, such as health clinics, schools, community-owned facilities, religious structures and cemeteries, which must be replaced.

Budgets

Funding is also a critical component of the relocation institutional design and implementation. Relocation should not be regarded as a goodwill gesture towards people facing climate change-induced displacement. Affected people should receive assistance in relocation as a matter of right. Successful relocation requires substantial budgetary resources to meet all costs of implementing the relocation plan. National and regional government agencies need to prepare budgets and make arrangements to ensure the timely flow of funds for relocation.

Principle 13 emphasizes the importance of a budget, by outlining the multi-level investment that needs to occur so that 'all appropriate administrative, legislative and judicial measures, including the creation of adequately funded Ministries, departments, offices and/or agencies at the local (in particular), regional and national levels [are] empowered to develop, establish and implement an institutional framework'.

Replacement cost addresses compensation for tangible assets, primarily land, houses and other structures, access to water and electricity, and also the need for socio-economic support, such as training opportunities.

Conclusion

Relocation of populations will be a critical adaptation strategy in response to climate-induced environmental change. The Peninsula Principles outline the components of an adaptive relocation governance framework, based in human rights doctrine. A rights-based approach to relocation is essential in order to minimize and prevent the impoverishment and social disintegration that have plagued relocations mandated by government actors for geopolitical or infrastructure development purposes.

Note

1 The UN Declaration on the Rights of Indigenous Peoples affirms the right of Indigenous communities to make collective decisions affecting their fundamental human rights. In addition, Article 1 of the International Covenant on Civil and Political Rights (1966) specifically establishes that 'all peoples have the right of self-determination' by virtue of which 'they freely determine their political status and freely pursue their economic, social and cultural development.'

References

Abhas, K.J., 2010. *Safer Homes, Stronger Communities: A Handbook for Reconstructing after Natural Disasters.* Washington, DC: World Bank.

Alfredsson, G., 2013. Good Governance in the Arctic. In N. Loucheva (ed.) *Polar Textbook II.* Copenhagen: Nordic Council of Ministries.

Armitage, D. and Plummer, R., 2010. Adapting and Transforming: Governance for Navigating Change. In D. Armitage and R. Plummer (eds) *Adaptive Capacity and Environmental Governance.* Heidelberg: Springer.

Berkes, F., 2009. Evolution of Co-Management: Role of Knowledge Generation, Bridging Organizations and Social Learning. *Journal of Environmental Management* 90: 1692–1702.

Boege, V., 2011. *Challenges and Pitfalls of Resettlement Measures: Experiences in the Pacific Region.* Bielefeld: Center on Migration, Citizenship and Development Working Paper Series.

Broderstad, E.G. and Dahl, J., 2004. Political Systems. In: N. Einarsson, J.N. Larsen, A. Nilsson and O. Young (eds) *Arctic Human Development Report.* Akureyri: Stefansson Arctic Institute.

Bronen, R., 2010. Forced Migration of Alaskan Indigenous Communities Due to Climate Change. In T. Afifi and J. Jäger (eds) *Environment, Forced Migration and Social Vulnerability.* London and New York: Springer Verlag.

Bronen, R., 2011. Climate-Induced Community Relocations: Creating an Adaptive Governance Framework Based in Human Rights Doctrine. *N.Y.U. Review of Law and Social Change* 35(2): 101–148.

Bronen, R. and Chapin, F.S., 2013. Adaptive Governance and Institutional Strategies for Climate-induced Community Relocations in Alaska. *Proceedings of the National Academy of Sciences.* Washington, DC: National Academy of Sciences.

Brookings-Bern (Brookings-Bern Project on Internal Displacement), 2011. *IASC Operational Guidelines on the Protection of Persons in Situations of Natural Disasters.* Washington, DC: Brookings-Bern Project on Internal Displacement.

Brubaker, M.Y., Bell, J.N., Berner, J.E. and Warren, J.A., 2011. Climate Change Health Assessment: A Novel Approach for Alaska Native Communities. *International Journal of Circumpolar Health* 70(3): 266–273.

COHRE (Centre on Housing, Rights and Evictions), 2007. The Pinheiro Principles: Center on Housing, Rights & Evictions. Available from: www.unhcr.org.ua/img/uploads/docs/PinheiroPrinciples.pdf, accessed 12 May 2015.

Correa, E., 2011. Resettlement as a Disaster Risk Reduction Measure: Case Studies. In E. Correa (ed.) *Preventive Resettlement of Populations at Risk of Disaster: Experiences from Latin America.* Washington, DC: World Bank.

Emerson, K., Nabatchi, T. and Balogh, S., 2011. An Integrated Framework for Collaborative Governance. *Journal of Public Administration Research and Theory Advance Access* 22(1): 1–29.

Ferris, E., 2012. *Protection and Planned Relocations in the Context of Climate Change*. Geneva: UN High Commission of Refugees, Division of International Protection.

Grannis, J., 2011. *Adaptation Tool Kit: Sea-Level Rise and Coastal Land Use* Washington, DC: Georgetown Climate Center.

Holthus, P., Crawford, M., Makroro, C. and Sullivan, S., 1992 *Vulnerability Assessment for Accelerated Sea Level Rise – A Case Study: Majuro Atoll, Republic of the Marshall Islands*, Samoa: South Pacific Environment Program Reports and Study Series 60.

IPCC, 2012. Summary for Policymakers. In C.B. Field, V. Barros, T.F. Stocker, D. Qin, D.J. Dokken, K.L. Ebi, M.D. Mastrandrea, K.J. Mach, G.-K. Plattner, S.K. Allen, M. Tignor and P.M. Midgley (eds) *Managing the Risks of Extreme Events and Disasters to Advance Climate Change Adaptation: A Special Report of Working Groups I and II of the Intergovernmental Panel on Climate Change*. Cambridge, UK, and New York: Cambridge University Press, pp. 3–21.

Kannen, A. and Forbes, D.L. 2011. Integrated Assessment and Response to Arctic Coastal Change. In D.L. Forbes (ed.) *State of the Arctic Coast 2010 – Scientific Review and Outlook*. Geesthacht: International Arctic Science Committee, Land-Ocean Interactions in the Coastal Zone, Arctic Monitoring and Assessment Programme, International Permafrost Association, http://arcticcoasts.org, accessed 22 May 2015.

Kofinas, G.P., 2009. Adaptive Co-management in Social-Ecological Governance. In F.S. Chapin, III, G.P. Kofinas and C. Folke (eds) *Principles of Ecosystem Stewardship: Resilience-Based Natural Resource Management in a Changing World*. New York: Springer.

Leckie, S., 2009. Climate-related Disasters and Displacement: Homes for Lost Homes, Lands for Lost Lands. In J.M. Guzmán, G. Martine, G. McGranahan, D. Schensul and C. Tacoli (eds) *Population Dynamics and Climate Change*. London and New York: International Institute for Environment and Development and UN Population Fund. pp. 119–132.

Lewis, D.A., 2012. The Relocation of Development from Coastal Hazards through Publicly Funded Acquisition Programs: Examples and Lessons from the Gulf Coast. *Sea Grant Law and Policy Journal* 5(1): 98–139.

May, B. and Plummer, R., 2011. Accommodating the Challenges of Climate Change Adaptation and Governance in Conventional Risk Management: Adaptive Collaborative Risk Management (ACRM). *Ecology and Society* 16(1): 47.

Moyn, S., 2010. *The Last Utopia, Human Rights in History*. Cambridge, MA: Harvard University Press.

Office of the United Nations High Commissioner of Human Rights, 2007. *Consultation on Human Rights and Access to Safe Drinking Water and Sanitation* 4. Geneva: UNHCR.

Office of the United Nations High Commissioner on Human Rights, 2008. *Consultation on the Relationship Between Climate Change and Human Rights* 4. Geneva: UNHCR.

Risse, M., 2009. The Right to Relocation, *Ethics and International Affairs* 23: 281.

Sallenger, A.H. Jr., Doran, K.S. and Howd, P.A., 2012. Hotspot of Accelerated Sea-Level Rise on the Atlantic Coast of North America. *Nature Climate Change* 2: 884–888.

Sen, A., 2004. Elements of a Theory of Human Rights. *Philosophy and Public Affairs* 32.

Smith, A.O., 2009. Introduction In A.O. Smith, (ed.) *Development and Dispossession: The Crisis of Forced Displacement and Resettlement*, Santa Fe, NM: School for Advanced Research Press.

UN, 1948. Universal Declaration of Human Rights. New York: United Nations General Assembly.

UN, 1951. UN Convention Relating to the Status of Refugees. New York: United Nations General Assembly.

UN, 1967. Protocol Relating to the Status of Refugees, New York: United Nations General Assembly.

UN, 1992. Framework Convention on Climate Change. New York: United Nations General Assembly.

UN, 2006. Convention for the Safeguarding of the Intangible Cultural Heritage. New York: United Nations General Assembly.

UN, 2007. Universal Declaration on the Rights of Indigenous Peoples. New York: United Nations General Assembly,.

UN Committee on Economic, Social and Cultural Rights, 1991. General Comment No. 4: The Right to Adequate Housing (Sixth session, 1991), UN Doc. E/1992/23, annex III at 114 (1991), reprinted in Compilation of General Comments and General Recommendations Adopted by Human Rights Treaty Bodies, UN Doc. HRI/GEN/1/Rev.6 at 18 (2003).

UN Committee on Economic, Social and Cultural Rights, 1999. General Comment No. 12: The Right to Adequate Food, 1 UN Doc. E/C.12/1995/5 (12 May 1999).

UN Committee on Economic, Social and Cultural Rights, 2006. Fact Sheet No. 16 (Rev. 1) 4, New York.

UN Human Rights Committee, 1982. General Comment No. 6: The Right to Life, UN Doc. HRI/GEN/1/Rev.7 (30 April 1982).

UN Human Rights Committee, 2011. *The Right to Food*, A/HRC/RES/16/27, paragraph 11.

Winkler, M.S., Krieger, G.R., Divall, M.J., Cissé, G., Wielga, M., Singer, B.H., Tanner, M. and Utzinger, J. 2013. Untapped Potential of Health Impact Assessment. *Bulletin of the World Health Organization* 91(4): 298–305.

Woodward-Clyde Consultants, 1998. *Understanding and Evaluating Erosion Problems*. Anchorage, AK: Department of Community and Regional Affairs.

World Bank, 2004. *Involuntary Resettlement Sourcebook: Planning and Implementation in Development Projects*. Washington, DC: World Bank.

7 The responsibilities of states to protect climate-displaced persons

Ezekiel Simperingham

Introduction

Section IV of the Peninsula Principles, entitled 'Displacement', contains three Principles relating to the responsibilities of States towards climate-displaced persons: the right to State assistance for climate-displaced persons; the right to housing and livelihoods; and the right to remedies and compensation.

This chapter addresses each of these Principles in turn, detailing the legal and normative basis for each, as well as providing commentary on relevant practice. The chapter also addresses the corresponding responsibilities of the international community, particularly where a State is unable or unwilling to meet its primary responsibility towards climate-displaced persons.

Each of the three Principles in Section IV addresses the responsibilities of States towards 'climate displaced persons experiencing displacement but who have not yet been relocated' (hereinafter 'climate-displaced persons'). This category of climate-displaced persons is more restricted than the general category of 'climate-displaced persons' that applies elsewhere in the Peninsula Principles.[1] In many ways, the focus on 'climate-displaced persons experiencing displacement, but who have not yet been relocated', mirrors the definition of internally displaced persons (IDPs) contained in the Guiding Principles on Internal Displacement (the Guiding Principles).[2] However, a key distinction is that the Peninsula Principles clearly apply to persons displaced by slow-onset as well as sudden-onset disasters.[3] It is less clear whether the Guiding Principles apply to persons displaced as a result of slow-onset environmental events,[4] although recently it has been claimed that the Guiding Principles may be flexible enough to include such events.[5]

Principle 14: State assistance to those climate displaced persons experiencing displacement but who have not been relocated

> Principle 14a States have the primary obligation to provide all necessary legal, economic, social and other forms of protection and assistance to those climate displaced persons experiencing displacement but who have not been relocated.

Principle 14a emphasizes that States have the 'primary obligation' to provide all necessary legal, economic, social and other forms of protection and assistance to climate-displaced persons. This is based on the Principle in international law that providing protection and humanitarian assistance to nationals is a primary duty and responsibility of the State.[6]

However, where States are unable to provide protection or assistance to their nationals, the role of the international community becomes 'of great importance'.[7] In such situations, the General Assembly has emphasized that States should 'facilitate the work of [intergovernmental and non-governmental] organizations in implementing humanitarian assistance'.[8] However, the General Assembly has also emphasized that humanitarian assistance 'should be provided with the consent of the affected country and in Principle on the basis of an appeal'.[9]

Where a State is unable or unwilling to meet its responsibilities to protect and assist its citizens, including climate-displaced persons, and there is no request or appeal to the international community for assistance, there is an arguable role for the international community to take action under the Responsibility to Protect (R2P). Specifically in the context of natural disasters, it has been argued that the R2P could arise where 'the disaster may have begun as a natural disaster but it quickly turned into a human-made disaster in which crimes – that could well constitute crimes against humanity – were committed'.[10]

However, there have also been a number of strong commentaries claiming that the R2P does not currently extend to natural disasters or climate change, including the Report of the Secretary General on Implementing the Responsibility to Protect, which stated:

> The [R2P] applies, until Member States decide otherwise, only to the four specified crimes and violations: genocide, war crimes, ethnic cleansing and crimes against humanity. To try to extend it to cover other calamities, such as HIV/AIDS, climate change or the response to natural disasters, would undermine the 2005 consensus and stretch the concept beyond recognition or operational utility.[11]

Furthermore, even if the response or lack of response to a climate change-linked natural disaster could amount to a crime against humanity, it is not clear what action (based on the R2P) the international community could or could not legitimately take. At a minimum, any international action would require a meeting of the Security Council to consider what steps to take.

Principle 14a also emphasizes that climate-displaced persons must be recognized and treated as persons entitled to enjoy the same rights as others in their country, including the right to protection and assistance by the State.[12] Climate-displaced persons must not be discriminated against on any basis, including their displacement, race, colour, sex, disability, language, religion, political opinion, national or social origin, property, birth, age or any other status.[13]

However, as climate-displaced persons may have specific vulnerabilities and needs distinct from those of the non-affected population, States should provide for specific and targeted assistance and protection measures,[14] including undertaking all necessary legal, economic, social and other forms of protection and assistance required to ensure that all human rights of climate-displaced persons are respected, protected and fulfilled.

> Principle 14b Protection and assistance activities undertaken by States should be carried out in a manner that respects both the cultural sensitivities prevailing in the affected area and the Principles of maintaining family and community cohesion.

The responsibility of States to undertake protection and assistance includes the responsibility to be 'respectful of the culture of individuals, minorities, peoples and communities'[15] and ensure that the provision of humanitarian goods and services, for example food, medicine and clothing, are culturally acceptable to affected persons, particularly if they are members of indigenous peoples or belong to particular ethnic or religious communities.[16]

The humanitarian response to the Pakistan floods of 2010 provides an example of the challenge of ensuring culturally acceptable protection and assistance in a humanitarian context. In that response it was clear that seeking aid was especially difficult for some women in areas with cultural norms that placed shame upon receiving aid or medical care from a male.[17] It was thus essential during the humanitarian response to ensure that there were sufficient numbers of trained female aid workers to provide all affected women with assistance and protection.

In the context of food assistance to climate-displaced persons, it should be ensured that food aid commodities are culturally acceptable and that the affected population has the knowledge and means to process and prepare the foods using their usual cooking facilities and fuel. It has been emphasized that 'emergencies are not a suitable time to introduce new types of food'.[18] One potential solution to ensuring culturally appropriate food assistance is to provide affected populations with cash transfers or vouchers to purchase food locally (and where available) rather than providing predetermined and potentially unsuitable food aid commodities.

Ensuring the 'adequacy' of shelter and housing for climate-displaced persons requires that any shelter or housing is 'culturally adequate',[19] meaning that 'the way housing is constructed, the building materials used and the policies supporting these must appropriately enable the expression of cultural identity and diversity of housing.'[20] In some contexts, it may be more effective to ensure culturally appropriate shelter or housing by providing affected populations with cash transfers, vouchers or materials to repair or construct their home, rather than providing them with a complete and potentially unsuitable house or shelter. For example, in the humanitarian response to Typhoon Bopha in the Philippines, where approximately 60–80 per cent of the affected population were from an indigenous group, a number of indigenous communities expressed concern that communal shelter designs were not appropriate for indigenous people, who were accustomed to living in one house per family. The Philippines National Commission on Indigenous Communities emphasized that indigenous persons would prefer the provision of materials to rebuild their own homes.[21]

However, States must equally ensure that protection and assistance activities, while respecting the cultural sensitivities prevailing in an affected area, do not also contravene human rights standards.[22]

Principle 14b also emphasizes that States must act in a manner that ensures the Principles of maintaining family and community cohesion. Experience has shown that families and communities are frequently separated as a result of displacement.[23] Such separation may occur deliberately where parents entrust their children to the care of others, or accidentally, during flight or when seeking shelter in large and crowded camps.[24] Separation may also occur through humanitarian interventions, for example when shelter is provided for some, but not all, family members, or during evacuations from danger zones.[25]

Human rights law requires States to ensure protection and assistance to the family and to ensure freedom from arbitrary or unlawful interference with the family.[26] This responsibility requires States to undertake any necessary legislative, administrative or other measures to prevent family separation.[27] In the context of displacement, this means that 'families who wish to remain together should be allowed and assisted to do so ... and their separation should be prevented'.[28] At all times, States and humanitarian actors must act to ensure that children are not separated from their parents against their will or against their best interests.[29]

Where family separation occurs, States have the obligation to facilitate family reunion – this is addressed under Principle 14cix:

> Principle 14c States should provide climate displaced persons experiencing displacement but who have not been relocated with a practicable level of age and gender-sensitive humanitarian assistance.

The responsibility to provide age- and gender-sensitive humanitarian assistance is part of the broader obligation to ensure that no discrimination occurs in the provision of humanitarian assistance. Children, adolescents, older persons and women are all potentially vulnerable groups in the context of climate displacement.

In addition to the risk of family separation (addressed above), climate-displaced children are exposed to the risk of exploitation, including forced recruitment, abduction, trafficking or sexual exploitation.[30] Climate-displaced adolescents may also be vulnerable to sexual abuse, exploitation and forced labour.[31] Older persons may suffer social and economic hardship where they become separated from their families and other support structures. They may also be at increased risk of violence, exploitation or abuse.[32]

Displaced women and girls, in addition to the risks associated with family separation (addressed above), often enjoy even less social, economic and political equality and are less represented in formal leadership structures than men.[33] Women in emergencies can face a heightened risk of economic and sexual exploitation, including trafficking and contemporary forms of slavery.[34] Note, however, that gender-sensitive humanitarian assistance should focus not only on the particular protection risks that women and girls face but also those faced by men and boys.[35]

Targeted measures should be undertaken to address the specific protection and assistance needs of children, adolescents, older persons and women (as well as men and boys).[36] Such needs should be identified and assessed on the basis of non-discriminatory and objective criteria, and in consultation with the affected population.[37] In all decisions and actions concerning children, the best interests of the child should be a primary consideration.[38]

All climate-displaced persons should be protected against gender-based violence (GBV) and survivors of such violence should be provided with appropriate support.[39] National authorities have the responsibility to prevent and respond to GBV, which includes taking all necessary legislative, administrative, judicial and other measures to prevent, investigate and punish acts of GBV and to ensure adequate care, treatment and support to survivors of GBV, based on the Principles of respect, confidentiality, safety, security and non-discrimination.[40]

The provision of humanitarian goods and services should also be 'sensitive to gender and age requirements'.[41] For example, in any temporary or collective shelters, it is essential to ensure that water, sanitation and hygiene (WASH) facilities are within easy and safe access and are private, in order to ensure that women and girls are comfortable to use the facilities during both day and night hours and to avoid the risk of women and girls walking to remote locations to collect water for drinking, cooking or laundry, putting them at risk of harassment and sexual assault.[42]

Principles 14ci–ix of the Peninsula Principles highlight a number of specific areas of humanitarian assistance that States are responsible for providing to climate-displaced persons. Principle 14c makes clear that this list of areas of humanitarian assistance is not exhaustive: for example, particular communities may have specific needs based on livelihood strategies or cultural preferences.

Principle 14ci: Emergency humanitarian services

Principle 14ci emphasizes that States should provide climate-displaced persons with emergency humanitarian services. Services typically required during the emergency phase of a humanitarian response include all life-saving measures, including food, water and sanitation, shelter, clothing and health services. The obligation to provide emergency humanitarian services is closely related to the primary obligation of States to protect displaced persons (discussed above under Principle 14a). However, as discussed, where the State is unable or unwilling to provide emergency humanitarian services, the role of the international community becomes important, especially where the State has requested the assistance of the international community in meeting its humanitarian obligations towards climate-displaced persons.

It is important to emphasize the responsibility of States not only to provide humanitarian goods and services to climate-displaced persons, but equally to provide 'adequate' goods and services. This means that goods and services must be available, accessible, acceptable and adaptable:[43]

- *Availability* means that goods and services must be provided in sufficient quantity and quality. For example, the provision of food to climate-displaced persons will only be sufficient if the correct amounts of energy and nutrition are provided and the foods are acceptable and familiar to the population.[44]
- *Accessibility* requires that goods and services are provided according to needs and without discrimination; that they are within safe reach and can be physically accessed by everyone, including persons with specific needs, and are known to the beneficiaries. For example, during the construction of temporary shelters (known as 'bunkhouses') during the humanitarian response to Typhoon Haiyan in the Philippines, it was emphasized that each bunkhouse needed to meet minimum accessibility requirements, so that the temporary shelters were accessible to all persons, including the elderly and those with disabilities.[45]
- *Acceptability* refers to the requirement that goods and services provided are respectful of the culture of individuals, minorities, peoples and communities, and sensitive to gender and age requirements.
- *Adaptability* requires that these goods and services be provided in ways flexible enough to adapt to the change of needs in the different phases of emergency relief, recovery and, in the case of internally displaced persons (including climate-displaced persons), return, local

integration or settlement elsewhere in the country.[46] For example, in the early stages of a humanitarian response it may be most effective to support affected populations with a 'core' house or shelter,[47] designed with the expectation that the homeowner will add on to the structure when circumstances permit. For displaced persons living in temporary shelters, who are expected to return to their original home or relocate elsewhere, it is essential to provide goods and services that are not only portable where possible, but also where ownership has been clarified, for example regarding ongoing ownership of any tents and tarpaulins that have been provided to the affected population.

Actors contributing to the humanitarian response should strive to achieve all elements of these criteria as soon as feasible. However, during the emergency phase, food, water and sanitation, shelter, clothing and health services are considered adequate if they respond to what is needed for survival and meet internationally recognized standards, including Sphere Standards.[48]

Principle 14cii: Evacuation and temporary and effective permanent relocation

Where natural disasters create imminent risks to individuals and communities, States should take measures to ensure that those persons are protected and able to remain wherever they may be located.[49] However, if such measures are not sufficient, the departure of persons from areas of imminent risk should be facilitated.[50] Where endangered persons cannot leave on their own, they should be evacuated.[51]

Persons should not be evacuated against their will, unless the following conditions are met: the forced evacuation is provided for by law; the evacuation is absolutely necessary to respond to a serious and imminent threat to life or health, and less intrusive measures would be insufficient to avert that threat and, to the extent possible, the persons concerned have been informed and consulted about the forced evacuation.[52] Evacuations, whether voluntary or forced, should be carried out in a manner that fully respects the rights to life, dignity, liberty and security of those affected and that does not discriminate.[53] Persons who leave or are evacuated should be supported to stay as close to their places of habitual residence as the security and safety situation allows.[54] The designated evacuation centres or temporary shelter zones should be safe and not expose them to further risk.[55]

After the emergency phase of the humanitarian response, climate-displaced persons should be granted the opportunity to return, locally integrate or settle elsewhere in the country.[56] In some situations, States may wish to restrict the ability of affected persons to return to their homes or places of former habitual residence. For example, in the weeks after Typhoon Haiyan devastated the central Philippines, a '40-metre no-build zone' was declared along the coastline of the affected regions.[57] This had the impact of restricting the ability of displaced persons to return and rebuild their homes if they were in the designated no-build zone. The stated intention was that all persons who previously resided in the no-build zone would be relocated to safe areas. However, a number of complications and concerns became apparent, including the arbitrary basis of the 40-metre no-build zone and whether there had been sufficient consideration of alternatives that would allow people to return safely, for example undertaking adaptive measures such as the planting of protective mangroves or the building of raised houses on stilts. It also became clear that there was insufficient land or resources to effectively relocate all persons living in the no-build zone in a manner that fully respected their rights.

Any permanent prohibition on return, without the consent of affected persons and communities, such as the imposition of the no-build zone in the Philippines or other forced relocations, should only be considered and implemented if the area where people live or want to return to is an area with high and persistent risk for life and security that cannot be mitigated by available adaptation and other protective measures.[58] As with forced evacuations, any such restrictions must also be: provided for by law; necessary and solely implemented to protect the lives and health of the affected population; and only imposed where the risks to lives and health could not be mitigated by other adaptation or less intrusive protective measures.[59]

International standards are also clear that involuntary relocation must be viewed as a measure of last resort.[60] In all cases of prohibitions on remaining, returning and rebuilding, measures should be taken to provide owners with due process guarantees, including the right to be heard and the right of access to an independent court or tribunal, as well as just compensation.[61]

Where relocation is not voluntary, international standards on forced evictions must be adhered to. Forced evictions are recognized under international law as being *prima facie* incompatible with the right to adequate housing and can 'only be justified in the most exceptional circumstances, and in accordance with the relevant Principles of international law'.[62] Whether or not the relocation is voluntary, it should meet international standards and best practice guidelines, including ensuring adequate site selection, shelter and WASH and other facilities at the relocation site. Affected communities should also have sufficient information and be adequately consulted in any relocation planning. Finally, non-discrimination should be ensured throughout the relocation process and the rights of the most vulnerable must be protected.[63]

Principle 14ciii: Medical assistance and other health services

The responsibility to provide medical assistance and other health services to climate-displaced persons forms part of the broader responsibility of

States to respect, protect and fulfil the right to an adequate standard of living and the right to health. The responsibility to ensure an adequate standard of living includes providing and facilitating access to essential medical services.[64]

Displacement can profoundly impact the health and well-being of individuals and communities. A lack of access to adequate shelter, sanitation, food and safe water can undermine people's ability to prevent and respond to health-related risks in their environment.[65] These health-related risks are often compounded by limited access to health-care facilities, services and supplies during displacement.[66] When designing and planning temporary shelter or collective centres, health factors (particularly related to water, sanitation and hygiene) are a key concern for the affected population. This is especially the case where people are living in buildings never designed for even temporary habitation (for example, the use of sports stadiums as evacuation centres following Hurricane Katrina in New Orleans and Typhoon Haiyan in Tacloban). Inadequate conditions in collective centres can have serious health repercussions, including on psychosocial health.[67]

The responsibility on States to ensure the highest attainable standard of health, including for climate-displaced persons, is contained in the Universal Declaration of Human Rights,[68] the International Convention on the Elimination of All Forms of Racial Discrimination,[69] the International Covenant on Economic, Social and Cultural Rights[70] and the Convention on the Rights of the Child,[71] and has been further clarified by the UN Committee on Economic, Social and Cultural Rights.[72]

In the context of displacement, the Guiding Principles on Internal Displacement emphasize that 'regardless of the circumstances, and without discrimination, competent authorities shall provide internally displaced persons with and ensure safe access to ... essential medical services and sanitation'.[73] In practice, national authorities should also seek to improve health and hygiene awareness among affected populations through awareness-raising campaigns and education.

Principle 14civ: Shelter

The responsibility of States to ensure adequate shelter for all persons during the emergency phase of a humanitarian response is an aspect of the broader obligation to ensure the right to an adequate standard of living and, in particular, an adequate standard of housing (addressed under Principle 15 below).

During the initial stages of displacement following a disaster, ensuring adequate shelter is essential to survival and is necessary to ensure security, personal safety and protection from the climate. Shelter assists in ensuring human dignity, sustaining family and community life and enabling affected populations to recover from the impacts of a disaster.[74] The shelter standards that apply during the emergency humanitarian response reflect the core content of the right to adequate housing, and are more limited than the complete expression of the right. The Sphere Standards provide useful guidance on the core content of the right to adequate housing[75] and highlight five elements that apply in an emergency situation: strategic planning; settlement planning; covered living space; construction; and environmental impact.[76]

The Sphere Standards also highlight that the meaningful participation of the community in decisions related to shelter is essential; that groups with specific needs require additional attention to ensure their access to adequate shelter, including displaced unaccompanied children, single women or female-headed households, unaccompanied older persons, older persons heading households, disabled persons, ethnic or religious minority groups and displaced families who do not own land or property;[77] and that gradual improvement of shelter throughout the displacement cycle is vital in order to contribute to a durable solution after displacement ends.

Further, camps and collective centres should be established as a last resort for climate-displaced persons and only where the possibility of host family arrangements, self-sustainability or rapid rehabilitation does not exist.[78] However, a concern is that considerably less is known about displaced persons living with host families compared to those living in camps and collective centres. This is partially a matter of practicality – it is easier to monitor a group of people registered and living in the same location compared to disparate persons throughout the community – but also, historically, it was assumed that displaced persons in hosting arrangements were located mainly in urban areas which were likely to be more affluent than the camps, and they were consequently less vulnerable or in need of assistance. Although host arrangements may be preferable, they also require monitoring and the provision of support where required, particularly as host families and host communities may be under economic strain from sharing meagre resources with displaced persons. Examples of assistance that can be provided to displaced persons in host arrangements and their hosts include cash-based assistance, community infrastructure, support to information or social development centres and quick impact projects (QIPs). It is also important to note that supporting hosting situations may be more complex and time-consuming than assisting displaced persons in camps and collective centres; and that in some cases camps may provide greater protection from violence and abuse. Despite this, the aim must be to respect how displaced persons choose to cope with their displacement and to support their choices where possible.[79]

Where States do not have the ability or resources to ensure adequate shelter during the emergency humanitarian phase, they should request international cooperation and assistance to ensure that their primary obligation to ensure the right to adequate shelter is met for all persons.[80]

Principle 14cv: Food

Access to adequate food and nutrition is one of the most important priorities during the emergency humanitarian phase. During displacement, all stages of production, procurement, preparation, allocation and consumption of food may be disrupted, often with impacts on health and nutrition.[81] States have the primary responsibility to ensure the right to adequate food for all persons, including climate-displaced persons.[82] The Guiding Principles establish that 'at the minimum, regardless of the circumstances and without discrimination, competent authorities shall provide IDPs, including climate-displaced persons with, and ensure safe access to, essential food.'[83] As an alternative or perhaps complement to direct food aid, the right to adequate food and nutrition may also be realized through cash transfers. In assessing whether cash or food transfers may be a more appropriate option in a given circumstance, it is important to take into account the broader context around the affected population, including the economics of food consumption, market analysis, cost effectiveness and efficiency, capacity requirements and the preferences of the displaced persons themselves.[84]

Where States are unable to ensure the right to adequate food, they should request international assistance and cooperation to ensure that their primary obligation is met. States should also facilitate safe and unimpeded access for international food-related assistance.[85] States have the obligation to avoid discrimination in access to adequate food, including discrimination on the basis of displacement.[86]

Principle 14cvi: Potable water, and Principle 14cvii: Sanitation

The human right to water entitles everyone to sufficient, safe, acceptable, physically accessible and affordable water for personal and domestic uses.[87] The responsibility of States to ensure the right to water is an aspect of the right to an adequate standard of living and the right to the highest attainable standard of health.[88]

In the context of displacement, States must take steps to ensure the availability of a sufficient and constant supply of water of adequate quality. Water and sanitation facilities must also be accessible to everyone without discrimination of any kind. This includes ensuring that such facilities are located within the safe reach of all sections of the population, designed in an age- and gender-sensitive way, and affordable to all.[89] During displacement, States must assist populations to urgently secure access to water, sanitation and basic hygiene facilities to ensure their survival, especially for protection against diseases and for their dignity, until they can return home or find another durable solution.[90] Displaced persons must have access to adequate water whether they stay in camps or in urban and rural areas.[91]

Access to water has also been a source of conflict within and between communities. It is essential properly to plan and implement conflict-sensitive WASH programmes to reduce tensions between communities.[92]

> Principle 14cviii measures necessary for social and economic inclusion including, without limitation, anti-poverty measures, free and compulsory education, training and skills development, and work and livelihood options, and issuance and replacement of lost personal documentation.

Principle 14cviii provides a non-exhaustive list of measures necessary for the social and economic inclusion of climate-displaced persons, including free and compulsory education, training and skills development, and work and livelihood options and issuance and replacement of lost documentation.

Human rights law guarantees the right to free and compulsory education for all at the primary level.[93] The development of accessible and affordable secondary education, including both general and vocational education as well as higher education, is also encouraged.[94] Human rights law also prohibits any form of discrimination in the right to education, including on the grounds of being displaced.[95]

The Guiding Principles on Internal Displacement emphasize that 'every human being has the right to education' and that to give effect to this right for internally displaced persons, 'the authorities concerned shall ensure that such persons, in particular displaced children, receive education which shall be free and compulsory at the primary level. Education should respect their cultural identity, language and religion.'[96] The Guiding Principles further emphasize that 'special efforts should be made to ensure the full and equal participation of women and girls in educational programmes' and that 'education and training facilities shall be made available to internally displaced persons, in particular adolescents and women, whether or not living in camps, as soon as conditions permit'.[97]

States must ensure that education is physically and economically accessible to everyone, including climate-displaced persons. Education must be of acceptable quality and must be adaptable, meaning that it is flexible and responds to the best interests of each child or adult.[98]

During a humanitarian crisis, it can understandably be difficult to provide all children with the same education and facilities that they received previously. However, the need to prioritize the provision of a quality basic education during the emergency phase is increasingly considered a key pillar of any humanitarian response. Beyond ensuring the right of children to an education, a basic emergency education provides a large number of benefits. The establishment of child-friendly spaces (such as schools) is critical to the physical and emotional protection of children; schools provide a platform to teach children survival skills (for example, how to access potable water following a natural disaster, or the location of safe evacuation routes in the event of any future disaster); child-friendly spaces also provide mothers with a safe space to care for their younger children and a central hub for aid distribution or the provision of health services outside school hours; and parents whose children attend school are also able more freely to pursue their recovery and rebuild their livelihoods.[99]

In the context of humanitarian assistance, Principle 14cviii highlights the vulnerability of climate-displaced persons to the loss of livelihoods, both in the context of being displaced and, indeed, as a factor leading to displacement. Principle 14cviii also highlights that the loss of livelihoods can lead to social and economic exclusion for climate-displaced persons. The loss of livelihoods may also engender a number of critical protection risks.[100] Being able to generate an income and maintain a livelihood during displacement improves the quality of life for individuals and assists in avoiding protracted dependency and exposure to further discrimination and abuse.[101]

The responsibility to support climate-displaced persons through training and skills development and work and livelihood options, in the context of humanitarian assistance, forms part of the core content of the obligation of States to ensure the right to work. For the more complete expression of the responsibility of States to ensure the right to work for climate-displaced persons, see Principle 15b below.

Principle 14cviii also highlights that displacement can lead to the loss, damage or destruction of personal documentation.[102] Lack of documentation can have serious consequences for climate-displaced persons, including restricted freedom of movement, limited access to assistance and services, exposure to harassment or arbitrary arrest and detention and the risk of statelessness.[103] For example, the loss of documentation proving tenure for housing or land can severely impact the ability of a climate displaced person to receive shelter or housing support.

States are required to take all measures necessary to ensure the right of all persons to recognition and equal protection under the law.[104] This includes ensuring that all persons, including climate-displaced persons, have a legal identity, as well as the means to prove their identity.[105]

The Guiding Principles on Internal Displacement emphasize that States shall issue displaced persons with 'all documents necessary for the enjoyment and exercise of their legal rights, such as passports, personal identification documents, birth certificates and marriage certificates'.[106] The Guiding Principles further highlight that States shall 'facilitate the issuance of new documents or the replacement of documents lost in the course of displacement, without imposing unreasonable conditions, such as requiring the return to one's area of habitual residence in order to obtain these or other required documents'.[107]

Principle 14cix: Facilitation of family reunion

Principle 14cix addresses the responsibilities of States to take steps towards the reunification of climate-displaced families. The responsibilities of States to maintain family cohesion are addressed under Principles 14c above.

The Guiding Principles on Internal Displacement provide that 'families which are separated by displacement should be reunited as quickly as possible'[108] and that

> all appropriate steps shall be taken to expedite the reunion of such families, particularly when children are involved. The responsible authorities shall facilitate inquiries made by family members and encourage and cooperate with the work of humanitarian organizations engaged in the task of family reunification.[109]

States should facilitate enquiries made by family members and assist affected persons in learning about the fate and whereabouts of missing relatives. Next of kin should be informed on the progress of the investigation and results obtained through the use of tracing services or mechanisms.[110]

Separated and unaccompanied children should be taken care of until they can be reunited with their families. All interim care arrangements should be in the best interests of the child.[111] When children are separated from their caregivers during displacement they are left vulnerable to violence, economic and sexual exploitation and trafficking. Further, current practices to document separated children, especially in resource-poor countries, are often outdated, inefficient and paper-driven. As a result, ensuring families are reunited following displacement can be a time-consuming and challenging process. To counter these concerns a number of innovations to expedite the process have been introduced, including RapidFTR (family tracing and reunification), an open-source mobile phone application that helps humanitarian workers collect, sort and share information about unaccompanied and separated children in emergency situations so that they can be registered for care services and reunited with their families.[112]

Principle 15: Housing and livelihood

> 15a States should respect, protect and fulfil the right to adequate housing of climate displaced persons experiencing displacement but who have not been relocated, which includes accessibility, affordability, habitability, security of tenure, cultural adequacy, suitability of location, and non-discriminatory access to basic services (for example, health and education).

Principle 15a highlights the responsibility of States to respect, protect and fulfil the right to adequate housing for climate-displaced persons. This

Principle focuses on the more complete expression of the right to adequate housing, as compared to the core of the right, addressed under Principle 14civ above.

The right to adequate housing has been recognized in a number of international legal instruments, declarations and other documents.[113] The responsibility of States to respect, protect and fulfil the right to adequate housing has been further clarified in a number of General Comments at the international level.[114] In particular, the Committee on Economic, Social and Cultural Rights has stated that in order for particular forms of shelter to be considered 'adequate', the following criteria should be taken into account: legal security of tenure; availability of services; affordability; habitability; accessibility; location; and cultural adequacy.[115]

In the context of displacement, ensuring legal security of tenure can be a challenging component of the humanitarian response. It is important to emphasize that tenure can take a variety of forms beyond formal private ownership, for example rental (public and private) accommodation, cooperative housing, lease, owner-occupation, emergency housing and informal settlements, including occupation of land or property. Such tenure types are applicable in statutory, customary or religious systems. This means that in ensuring the right to housing for climate-displaced persons, national authorities, along with any humanitarian or human rights actors, should not strictly require formal private ownership over a particular plot of land as a precondition to shelter and housing support, but rather should take a broad and flexible approach to tenure that is based on the local legal and normative context and practice.

In addition to the above list of criteria for adequate housing, Principle 15a emphasizes that adequate housing must ensure 'non-discriminatory access to basic services (for example, health and education)'. This is based on the Principle that the right to adequate housing must not be subject to any form of discrimination,[116] as well as that the criterion 'location' (from the above list), requires that adequate housing 'must be in a location which allows access to employment options, health-care services, schools, childcare centres and other social facilities'.[117]

While the right to adequate housing applies to all persons, there are a number of specific provisions at the international level emphasizing its applicability to climate-displaced persons. The Guiding Principles on Internal Displacement provide that 'all internally displaced persons have the right to an adequate standard of living'[118] and that 'at the minimum, regardless of the circumstances, and without discrimination, competent authorities shall provide internally displaced persons with and ensure safe access to ... basic shelter and housing.'[119] The UN Principles on Housing and Property Restitution for Refugees and Displaced Persons (the Pinheiro Principles) emphasize that 'everyone has the right to adequate housing'[120] and that 'States should adopt positive measures aimed at alleviating the situation of refugees and displaced persons living in inadequate housing'.[121]

Principle 15 of the Peninsula Principles also highlights that States should respect, protect and fulfil the right to adequate housing. In the context of climate displacement, the responsibility to respect, protect and fulfil the right to adequate housing requires States to:

Respect the right to adequate housing by ensuring that no measures are taken which intentionally erode the legal and practical status of the right to adequate housing. States should comprehensively review relevant legislation, refrain from actively violating the right to adequate housing by strictly regulating forced evictions and ensure that the housing, land and property sectors are free from all forms of discrimination at all times. States should also assess national housing, land and property conditions, and accurately calculate, using statistical and other data and indicators, the true scale of non-enjoyment of the right to adequate housing, and the precise measures required for their remedy.[122]

Protect the right to adequate housing of climate-displaced persons by effectively preventing the denial of their rights by third parties and others capable of restricting these rights.[123]

Fulfil the right to adequate housing for climate-displaced persons, which involves issues of public expenditure, regulation of national economies and land markets, housing subsidy programmes, monitoring of rent levels and other housing costs, construction and financing of public housing, provision of basic social services, taxation, redistributive economic measures and any other positive initiatives that are likely to result in the continually expanding enjoyment of the right to adequate housing.[124]

> 15b Where climate displacement results in the inability of climate displaced persons to return to previous sources of livelihood, appropriate measures should be taken to ensure such livelihoods can be continued in a sustainable manner and will not result in further displacement, and opportunities created by such measures should be available without discrimination of any kind.

As discussed under Principle 14cviii, climate displacement leaves individuals and communities vulnerable to the loss of livelihoods, including the inability to return to previous sources of livelihood. For example, experts have warned that sea level rise in the Mekong Delta region may cause devastating floods, destroying crops and leading to the displacement of millions and the widespread loss of agricultural-based livelihoods.[125]

In these circumstances, Principle 15b emphasizes that States have the responsibility to take appropriate measures to ensure that such previously held livelihoods can be continued in a sustainable manner and will not result in further displacement. Principle 15b also emphasizes that opportunities created by such measures should be available without discrimination of any kind.

This responsibility is an aspect of State responsibility to ensure the right to work for all persons, including climate-displaced persons. The right to work is included in a number of human rights instruments and includes 'the right to free choice of employment and to just and favourable conditions of work and protection against unemployment';[126] 'the right to equal pay for equal work and to just and favourable remuneration'[127] and 'the right of everyone to the opportunity to gain his living by work which he freely chooses or accepts'.[128]

The International Covenant on Economic, Social and Cultural Rights states that in order to ensure the right to work, States should provide 'technical and vocational guidance and training programmes, policies and techniques to achieve steady economic, social and cultural development and full and productive employment under conditions safeguarding fundamental political and economic freedoms to the individual'.[129]

The Guiding Principles on Internal Displacement establish that displaced persons have 'the right to seek freely opportunities for employment and to participate in economic activities'.[130] However, as displaced persons often face more obstacles than others in achieving sustainable livelihoods and returning to previous sources of livelihoods, States should take specific measures to ensure the livelihoods of climate-displaced persons, including:[131]

- ensuring non-discriminatory access to public services, such as health, education, social welfare and housing loans to secure a stable and dignified environment so that displaced persons can become self-reliant;
- developing an appropriate legal and administrative framework to enable displaced persons to have access to the labour market while taking into account their specific needs, including recognition of academic and professional credentials, non-discriminatory job recruitment policies and psycho-social support;
- respecting and promoting the freedom of movement of displaced persons for economic purposes;
- integrating the development of displaced persons' livelihoods into national development programmes and poverty-reduction strategies; and
- securing all areas related to livelihoods and strengthening the rule of law to protect the assets and activities of displaced persons in their attempts to become self-reliant.
- Further livelihood support may include: facilitating access to savings and loan mechanisms; skills and vocational training; entrepreneurship training and business support services; training on the sustainable use of natural resources; labour-based activities; grant assistance in emergency situations; career guidance and support and facilitating access to apprenticeships and jobs as well as work and business permits.[132]

In the context of humanitarian assistance, the ability of States to provide training and skills development and work and livelihood options may well be limited. Where States are unable to meet their primary obligations towards climate-displaced persons, they should request the support and assistance of the international community. In responding to such a request, the international community should include early livelihood interventions as part of a multi-faceted strategy to reduce protection risks and further the realization of all human rights for climate-displaced persons. This is in contrast with the normal tendency for livelihood interventions to be focused on the post-humanitarian emergency phase.[133]

Principle 16: Remedies and compensation

> 16 Climate displaced persons experiencing displacement but who have not been relocated and whose rights have been violated shall have fair and equitable access to appropriate remedies and compensation.

Principle 16 emphasizes that where climate-displaced persons experience violations of their rights, States must ensure that they have fair and equitable access to appropriate remedies and compensation. This Principle is based on the general Principle under international law that all persons, including climate-displaced persons, have the right to an effective remedy for violations of their human rights and an enforceable right to compensation.

The right to an effective remedy is contained in the Universal Declaration of Human Rights,[134] the International Covenant on Civil and Political Rights,[135] the International Convention on the Elimination of All Forms of Racial Discrimination,[136] the Convention against Torture[137] and the Convention on the Rights of the Child.[138]

The Basic Principles and Guidelines on the Right to a Remedy and Reparation also highlight that States are required under international law to 'make available adequate, effective, prompt and appropriate remedies, including reparation'.[139]

In the context of climate displacement, experience from past environmental disasters has shown that individuals and communities are especially vulnerable to arbitrary or unlawful deprivation of their housing, land and property (HLP) rights. For example, property developers in Indonesia and Thailand moved in quickly after the 2004 tsunami, grabbing land from those relocated into resettlement camps to build luxury resorts.[140] Land grabbing also occurred after Hurricane Katrina in New Orleans, the Haiti earthquake, and cyclones and floods in the Philippines.[141]

The Pinheiro Principles directly address the right to a remedy and to compensation for displaced persons who have suffered violations of their HLP rights. Specifically, the Pinheiro Principles state that all

> displaced persons have the right to have restored to them any housing, land and/or property of which they were arbitrarily or unlawfully

deprived, or to be compensated for any housing, land and/or property that is factually impossible to restore as determined by an independent, impartial tribunal.[142]

In order to ensure effective remedies, States should ensure equal access to an effective judicial remedy. Victims should also have access to remedies through administrative and other bodies, as well as mechanisms, modalities and proceedings conducted in accordance with domestic law. The obligation to secure the right to access justice and fair and impartial proceedings must be ensured. To this end, States should:

- disseminate, through public and private mechanisms, information about all available remedies for gross violations of international human rights law;
- take measures to minimize the inconvenience to victims and their representatives, protect against unlawful interference with their privacy as appropriate, and ensure their safety from intimidation and retaliation, as well as that of their families and witnesses, before, during and after judicial, administrative or other proceedings that affect the interests of victims;
- provide proper assistance to victims seeking access to justice;
- make available all appropriate legal, diplomatic and consular means to ensure that victims can exercise their rights to remedy for gross violations of international human rights law.[143]

Conclusion

Section IV of the Peninsula Principles contains a broad range of rights to which climate-displaced persons are entitled, and to which States are under a duty to respect, protect and fulfil. These rights apply throughout the displacement cycle, from the initial emergency humanitarian phase, to situations of more protracted displacement, to the achievement of durable solutions.

States are under the duty to ensure that all of these rights apply to all persons equally, including to climate-displaced persons. However, as climate-displaced persons are often in a position of heightened vulnerability in regard to their basic rights, States must take special measures to ensure that their rights are protected. This chapter has provided examples of a number of practical measures that could be taken by States to ensure that the rights of all climate-displaced persons are respected, protected and fulfilled.

This chapter also addresses the important role of the international community in providing protection and assistance to climate-displaced persons, particularly where the State in question is unable or unwilling to meet its primary responsibility towards them.

In general, rather than requiring the development of considerable new normative standards for the protection of climate-displaced persons within States, Section IV of the Peninsula Principles demonstrates that a broad range of relevant human rights standards already exist and apply throughout all phases of the displacement cycle. Climate-displaced persons are already the holders of all of these rights and States already have the duty and primary responsibility to respect, protect and fulfil all of these rights. However, it is essential that States and climate-displaced persons and communities are made more aware of these rights and duties and that States, with the cooperation of the international community, take steps to respect, protect and fulfil all of these rights, based on best practice and genuine political will.

Notes

1 Principle 2c defines climate-displaced persons as 'individuals, households or communities who face or experience climate displacement' (The Peninsula Principles on Climate Displacement within States, August 2013).
2 UN Economic and Social Council (ECOSOC), Guiding Principles on Internal Displacement, 22 July 1998, E/CN.4/1998/53/Add.2.
3 Principle 2b of the Peninsula Principles defines climate displacement as the 'movement of people within a State due to the effects of climate change, including sudden and slow-onset environmental events and processes, occurring either alone or in combination with other factors'.
4 Principle 2 of the Guiding Principles on Internal Displacement defines IDPs as

> persons or groups of persons who have been forced or obliged to flee or to leave their homes or places of habitual residence, in particular as a result of or in order to avoid the effects of armed conflict, situations of generalized violence, violations of human rights or natural or human-made disasters, and who have not crossed an internationally recognized State border.

The focus on 'natural disasters' may exclude those individuals who are displaced by slow-onset environmental processes that do not qualify as a 'natural disaster'.
5 It has been stated that

> the addition of the words, 'in particular' [to the IDP definition] was intended to make the definition more elastic, enabling it to encompass new IDP categories that might arise but were unanticipated at the time the Principles were drafted. One such group was those uprooted by slow-onset natural disasters (e.g. drought, desertification, land degradation, rising sea levels). Unlike those uprooted by sudden-onset disasters (earthquakes, floods, hurricanes), they traditionally were seen as voluntary migrants (as opposed to forced migrants). However, the growing impact of climate change has begun to call this assumption into question and some, particularly those who can no longer return to their homes, are beginning to be seen as forced migrants or IDPs.
>
> (Roberta Cohen, *Lessons Learned from the Development of the Guiding Principles on Internal Displacement*, Georgetown University Working Paper, The Crisis Migration Project, Institute for the Study of International Migration, October 2013)

6 This Principle has been affirmed in a number of UN General Assembly Resolutions, see: General Assembly Resolutions 45/100 Humanitarian Assistance to Victims of Natural Disasters and Similar Emergency Situations, 14 December 1990; 46/182 Strengthening of the Coordination of Humanitarian Emergency Assistance of the United Nations, 19 December 1991; and 62/92 International Cooperation on Humanitarian Assistance in the Field of Natural Disasters, from Relief to Development, 1 February 2008.

7 The UN General Assembly has stated that 'the magnitude and duration of many emergencies may be beyond the response capacity of many affected countries. International cooperation to address emergency situations and to strengthen the response capacity of affected countries is thus of great importance' (General Assembly Resolution 46/182 'Strengthening of the Coordination of Humanitarian Emergency Assistance of the United Nations, 19 December 1991).

8 Ibid.

9 Ibid.

10 Roberta Cohen, 'The Burma Cyclone and the Responsibility to Protect: Speech', 21 July 2008.

11 UN General Assembly, 'Implementing the Responsibility to Protect: Report of the Secretary-General', 12 January 2009, A/63/677, paragraph 10(b).

12 Inter-Agency Standing Committee, Protecting Persons Affected by Natural Disasters. IASC Operational Guidelines on Human Rights and Natural Disasters, June 2006, p. 11.

13 Ibid.

14 Ibid., p. 10.

15 Brookings-Bern Project on Internal Displacement, IASC Operational Guidelines on the Protection of Persons in Situations of Natural Disaster, January 2011, p. 30, available at www.refworld.org/docid/4cd/e733be.html, accessed 13 May 2015.

16 Ibid., p. 31.

17 Reproductive Health Response in Crises Consortium, Needs of Women and Girls Must be Addressed in Pakistan Flood Response and Recovery, available at: http://rhrealitycheck.org/article/2010/08/06/meeting-reproductive-health-needs-women-girls-affected-pakistan-floods/, accessed 13 May 2015.

18 UNHCR, UNICEF, WFP and WHO, Food and Nutrition Needs in Emergencies, available at www.unscn.org/layout/modules/resources/handbook_2002_UNH.pdf, accessed 13 May 2015.

19 UN Committee on Economic, Social and Cultural Rights (CESCR), General Comment No. 4: The Right to Adequate Housing (Art. 11 (1) of the Covenant), 13 December 1991, E/1992/23.

20 Ibid., para 8(g).

21 Shelter Cluster, Legal and Regulatory Issues: Typhoon Bopha, 2013, available at www.sheltercentre.org/sites/default/files/regulatory_issues_in_bopha_response.pdf.

22 Brookings-Bern Project on Internal Displacement, IASC Operational Guidelines on the Protection Of persons in Situations of Natural Disaster, p. 12.

23 UNHCR, Handbook for the Protection of Internally Displaced Persons: Part V: Protection Risks: Prevention, Mitigation and Response. Action Sheet 9 – Family Unity, 1 December 2007, available at www.unhcr.org/4794bc2.html, accessed 13 May 2015.

24 Ibid.

25 Global Protection Cluster (GPC), *Handbook for the Protection of Internally Displaced Persons*, June 2010, p. 205, available at www.refworld.org/docid/4790cbc.html, accessed 13 May 2015.

26 Article 10, UN General Assembly, International Covenant on Economic, Social and Cultural Rights, 16 December 1966, United Nations, Treaty Series, vol. 993, p. 3; see also Article 16, UN General Assembly, Universal Declaration of Human Rights, 10 December 1948, 217 A (III); Articles 17 and 23 of the UN General Assembly, International Covenant on Civil and Political Rights, 16 December 1966, United Nations, Treaty Series, vol. 999, p. 171; Articles 16 and 18, UN General Assembly, Convention on the Rights of the Child, 20 November 1989, United Nations, *Treaty Series*, vol. 1577, p. 3.

27 GPC, *Handbook for the Protection of Internally Displaced Persons*, p. 207.

28 Brookings-Bern Project on Internal Displacement, IASC Operational Guidelines on the Protection of Persons in Situations of Natural Disaster, p. 49.

29 GPC, Handbook for the Protection of Internally Displaced Persons, p. 205.

30 Nils Kastberg, 'Strengthening the Response to Displaced Children', *Forced Migration Review*, Issue 15: Displaced children and adolescents: challenges and opportunities, October 2002, p. 4.

31 Internal Displacement Monitoring Centre (IDMC), Training on the Protection of IDPs Internally Displaced Children and Adolescents, available at http://idp-key-resources.org/documents/0000/d04382/000.pdf, accessed 13 May 2015.

32 GPC, *Handbook for the Protection of Internally Displaced Persons*, p. 13.

33 Ibid., p. 14.

34 Brookings-Bern Project on Internal Displacement, IASC Operational Guidelines on the Protection of Persons in Situations of Natural Disaster, p. 23.

35 GPC, *Handbook for the Protection of Internally Displaced Persons*, p. 14.

36 See Principle 4.2, ECOSOC, Guiding Principles on Internal Displacement.

37 Brookings-Bern Project on Internal Displacement, IASC Operational Guidelines on the Protection of Persons in Situations of Natural Disaster, p. 11.

38 UN General Assembly, Convention on the Rights of the Child.

39 Brookings-Bern Project on Internal Displacement, IASC Operational Guidelines on the Protection of Persons in Situations of Natural Disaster, p. 22.

40 GPC, *Handbook for the Protection of Internally Displaced Persons*, p. 172.

41 Brookings-Bern Project on Internal Displacement, IASC Operational Guidelines on the Protection of Persons in Situations of Natural Disaster, p. 30.

42 Sarah House, Suzanne Ferron, Marni Sommer and Sue Cavill, 'Violence, Gender and WASH: a Practitioners' Toolkit: Making Water, Sanitation and Hygiene Safer through Improved Programming', *Humanitarian Exchange Magazine*, Issue 60, February 2014, available at www.odihpn.org/humanitarian-exchange-magazine/issue-60/violence-gender-and-wash-a-practitioners-toolkit-making-water-sanitation-and-hygiene-safer-through-improved-programming, accessed 13 May 2015.

43 Brookings-Bern Project on Internal Displacement, IASC Operational Guidelines on the Protection of Persons in Situations of Natural Disaster, p. 30.

44 UNHCR, UNICEF, WFP and WHO, Food and Nutrition Needs in Emergencies.

45 CCCM Philippines, *Bunkhouse Assessment Report*, February 2014, available at http://cccmphilippines.iom.int/sites/default/files/documents/Bunkhouse%20Assessment%20Report%20Final%20Ver3_7%20Feb%202014.pdf, accessed 13 May 2015.

46 Brookings-Bern Project on Internal Displacement, IASC Operational Guidelines on the Protection of Persons in Situations of Natural Disaster, p. 29.

47 A core house is a small, well-constructed, durable and permanent structure that provides adequate living space and sanitation facilities. See for example, Habitat for Humanity, 'Core Houses', available at http://www.habitat.org/disaster/active_programs/Core_houses.aspx, accessed 13 May 2015.

48 The Sphere Project, *Humanitarian Charter and Minimum Standards in Disaster Response* (Geneva: The Sphere Project, 2011); Brookings-Bern Project on

Internal Displacement, IASC Operational Guidelines on the Protection of Persons in Situations of Natural Disaster, p. 30.

49 The IASC recommends, among other measures, informing persons about expected risks, proposed precautions and facilities such as safe escape routes and emergency shelters in their neighbourhoods; activating alert systems; integration of disaster awareness into educational curricula. See: Brookings-Bern Project on Internal Displacement, *IASC Operational Guidelines on the Protection of Persons in Situations of Natural Disaster*, p. 15.

50 Ibid., p. 15.

51 Ibid., p. 16.

52 Ibid., p. 16.

53 Ibid., p. 17.

54 Ibid.

55 Ibid., p. 18.

56 Brookings-Bern Project on Internal Displacement, IASC Framework on Durable Solutions for IDPs, April 2010, p. 5, available at www.refworld.org/docid/4c5149312. html, accessed 13 May 2015.

57 See for example, IRIN News, ' *"No-build zones" confusion delays resettlement of Haiyan survivors*', 18 July 2014, available at www.irinnews.org/report/100368/no-build-zones-confusion-delays-resettlement-of-haiyan-survivors.

58 Brookings-Bern Project on Internal Displacement, IASC Operational Guidelines on the Protection of Persons in Situations of Natural Disaster, p. 48.

59 Ibid., p. 16.

60 IFRC, Post-disaster Settlement Planning and Guidelines, 2012, p. 11, available at http://www.ifrc.org/PageFiles/71111/PostDisaster_Settlement_Guidelines.pdf, accessed 13 May 2015.

61 Inter-Agency Standing Committee, Protecting Persons Affected by Natural Disasters, para C.2.9.

62 CESCR, General Comment No. 4: The Right to Adequate Housing, para 18, and General Comment No. 7: The Right to Adequate Housing (Art.11.1): Forced Evictions, 20 May 1997, E/1998/22.

63 For further information, see IFRC, *Post-disaster Settlement Planning and Guidelines*, pp. 91–93, and Shelter Cluster, *Land Rights and Shelter: The Due Diligence Standard*, December 2013.

64 I Brookings-Bern Project on Internal Displacement, IASC Operational Guidelines on the Protection of Persons in Situations of Natural Disaster, p. 37.

65 GPC, *Handbook for the Protection of Internally Displaced Persons*, p. 264.

66 Ibid., p. 265.

67 Erin Mooney, 'When "Temporary" Lasts Too Long', *Forced Migration Review*, Issue 33, September 2009.

68 Article 25, UN General Assembly, Universal Declaration of Human Rights, 217 A (III).

69 Article 5(e)(iv), UN General Assembly, International Convention on the Elimination of All Forms of Racial Discrimination, 21 December 1965, United Nations, Treaty Series, vol. 660, p. 195.

70 Article 12, UN General Assembly, International Covenant on Economic, Social and Cultural Rights, p. 3.

71 Article 24.1, UN General Assembly, Convention on the Rights of the Child, p. 3.

72 CESCR, General Comment No. 14: The Right to the Highest Attainable Standard of Health (Art. 12 of the Covenant), 11 August 2000, E/C.12/2000/4.

73 Principle 18, ECOSOC, Guiding Principles on Internal Displacement.

74 Sphere Project, *Sphere Handbook: Humanitarian Charter and Minimum Standards in Disaster Response*, p. 244.

75 Ibid.

76 Ibid., pp. 249–267.

77 GPC, *Handbook for the Protection of Internally Displaced Persons*, p. 238.

78 Brookings-Bern Project on Internal Displacement, IASC Operational Guidelines on the Protection of Persons in Situations of Natural Disaster, p. 34; see also GPC, *Handbook for the Protection of Internally Displaced Persons*, p. 236.

79 UNHCR, *IDPs in Host Families and Host Communities: Assistance for Hosting Arrangements*, Geneva: UNHCR, 2012.

80 GPC, *Handbook for the Protection of Internally Displaced Persons*, pp. 238–239.

81 Sphere Project, *Sphere Handbook: Humanitarian Charter and Minimum Standards in Disaster Response*, p. 144.

82 GPC, *Handbook for the Protection of Internally Displaced Persons*, p. 248.

83 Guiding Principle 18.2, ECOSOC, Guiding Principles on Internal Displacement.

84 World Food Programme, *Cash and Food Transfers: A Primer*, Rome: WFP, 2007.

85 Sphere Project, *Sphere Handbook: Humanitarian Charter and Minimum Standards in Disaster Response*, p. 144.

86 GPC, *Handbook for the Protection of Internally Displaced Persons*, p. 253.

87 CESCR, General Comment No. 15: The Right to Water (Arts. 11 and 12 of the Covenant), 20 January 2003, E/C.12/2002/11.

88 See, for example, Article 25(1) of UN General Assembly, Universal Declaration of Human Rights, Article 14(2)(h) of UN General Assembly, Convention on the Elimination of All Forms of Discrimination Against Women, 18 December 1979, United Nations, Treaty Series, vol. 1249, p. 13; and Article 24(2)(c) of UN General Assembly, Convention on the Rights of the Child, p. 3.

89 GPC, *Handbook for the Protection of Internally Displaced Persons*, p. 257; note 'affordability' has been defined as meaning 'that direct, indirect, and other charges related to having water should be affordable to everyone' (Figaro Joseph, 'In Need of a Better WASH: Water, Sanitation, and Hygiene Policy Issues in Post-Earthquake Haiti', Oxfam America Research Backgrounder series, 2011, available at oxfamamerica.org/publications/wash-policy-issues-post-earthquake-haiti, accessed 13 May 2015.

90 GPC, *Handbook for the Protection of Internally Displaced Persons*, p. 256.

91 CESCR, General Comment No. 15: The Right to Water (Arts. 11 and 12 of the Covenant).

92 GPC, *Handbook for the Protection of Internally Displaced Persons*, p. 256.

93 See Article 26 of Universal Declaration of Human Rights, 10 December 1948, 217 A (III); Articles 24(2)(e) and (f), 28 and 29 of Convention on the Rights of the Child; Articles 6, 13 and 14 of UN General Assembly, International Covenant on Economic, Social and Cultural Rights, p. 3; Article 18 of UN General Assembly, International Covenant on Civil and Political Rights, p. 171; Article 5(e)(v) of UN General Assembly, International Convention on the Elimination of All Forms of Racial Discrimination, p. 195; Articles 10, 11(1)(c) and 14(2)(d) of UN General Assembly, Convention on the Elimination of All Forms of Discrimination Against Women, p. 13; UN Educational, Scientific and Cultural Organisation (UNESCO), Convention Against Discrimination in Education, 14 December 1960; and Principle 23(3)(4) of ECOSOC, Guiding Principles on Internal Displacement.

94 GPC, *Handbook for the Protection of Internally Displaced Persons*, p. 286.

95 Ibid.

96 Principle 23, UN Commission on Human Rights, Report of the Representative of the Secretary-General, Mr. Francis M. Deng, submitted pursuant to

Commission Resolution 1997/39. Addendum: Guiding Principles on Internal Displacement, 11 February 1998, E/CN.4/1998/53/Add.2.

97 Ibid.

98 See CESCR, General Comment No. 13: The Right to Education (Art. 13 of the Covenant); CESCR, General Comment No. 11: Plans of Action for Primary Education (Art. 14 of the Covenant), 10 May 1999, E/1992/23; UN Committee on the Rights of the Child (CRC), CRC General Comment No. 1: The Aims of Education, 17 April 2001, CRC/GC/2001/1, and GPC, *Handbook for the Protection of Internally Displaced Persons*, p. 286.

99 Humanitarian Coalition, *Education and Emergencies*, available at http://humanitariancoalition.ca/info-portal/factsheets/education-and-emergencies, accessed 13 May 2015.

100 For examples of the protection risks potentially triggered by the loss of livelihoods, see GPC, *Handbook for the Protection of Internally Displaced Persons*, p. 293.

101 Ibid., p. 292.

102 Ibid., p. 227.

103 Ibid., p. 228.

104 See Articles 6 and 7 of UN General Assembly, Universal Declaration of Human Rights, Article 26 of UN General Assembly, International Covenant on Civil and Political Rights, Article 15 of UN General Assembly, Convention on the Elimination of All Forms of Discrimination Against Women, Article 5(a) of UN General Assembly, International Convention on the Elimination of All Forms of Racial Discrimination, Articles 18 and 24 of UN General Assembly, International Convention on the Protection of the Rights of All Migrant Workers and Members of their Families, 18 December 1990, A/RES/45/158.

105 GPC, *Handbook for the Protection of Internally Displaced Persons*, p. 227.

106 Principle 20 ECOSOC, Guiding Principles on Internal Displacement.

107 Ibid.

108 Principle 17.3, ECOSOC, Guiding Principles on Internal Displacement.

109 Ibid.

110 Brookings-Bern Project on Internal Displacement, IASC Operational Guidelines on the Protection of Persons in Situations of Natural Disaster, p. 49.

111 Ibid.

112 For more information on RapidFTR, see: http://rapidftr.com.

113 Article 25(1) of UN General Assembly, Universal Declaration of Human Rights, Article 5(e)(iii) of UN General Assembly, International Convention on the Elimination of All Forms of Racial Discrimination, Article 27(3) of UN General Assembly, Convention on the Rights of the Child, Article 14(2) of UN General Assembly, Convention on the Elimination of All Forms of Discrimination Against Women, UN General Assembly, International Covenant on Civil and Political Rights, UN General Assembly, Convention Relating to the Status of Refugees, 28 July 1951, United Nations, Treaty Series, vol. 189, p. 137; International Labour Organization (ILO), Convention No. 169 Concerning Indigenous and Tribal Peoples; ILO, Convention No. 161 Concerning Occupational Health Services (1985); ILO, Convention No. 117 Concerning Social Policy (Basic Aims and Standards) (1962); ILO, Convention No. 110 Concerning Plantations (1958); ILO, Convention No. 82 Concerning Social Policy (Non-Metropolitan Territories) (1947); UN General Assembly, Declaration on the Right to Development, 4 December, 1986, A/RES/41/28; UN General Assembly Declaration on Social Progress and Development, 11 December 1969, A/RES/2542 (XXIV); UN General Assembly, Habitat: United Nations Conference on Human Settlements (1976); and the ILO Recommendation Concerning Workers' Housing (1961).

114 CESCR, General Comment No. 4: The Right to Adequate Housing (Art. 11 (1) of the Covenant), CESCR, General Comment No. 7: The Right to Adequate Housing (Art. 110 (1): forced evictions housing, 20 May 1997, E/1998/22.

115 CESCR, General Comment No. 4: The Right to Adequate Housing (Art. 11 (1) of the Covenant).

116 Ibid.

117 Ibid.

118 Principle 18.1, ECOSOC, Guiding Principles on Internal Displacement.

119 Principle 18.2(b), ECOSOC, Guiding Principles on Internal Displacement.

120 Principle 8.1, UN Sub-Commission on the Promotion and Protection of Human Rights, Principles on Housing and Property Restitution for Refugees and Displaced Persons, 28 June 2005, E/CN.4/Sub.2/2005/17.

121 Principle 8.2, ibid.

122 Displacement Solutions, *Climate Change Displaced Persons and Housing, Land and Property Rights: Preliminary Strategies for Rights-Based Planning and Programming to Resolve Climate Displacement*, p. 13.

123 Displacement Solutions, *Climate Change Displaced Persons and Housing, Land and Property Rights: Preliminary Strategies for Rights-Based Planning and Programming to Resolve Climate Displacement*, p. 13, available at http://displacementsolutions.org/files/documents/DS_Climate_change_strategies.pdf, accessed 13 May 2015.

124 Ibid.

125 IRIN News, 'Vietnam: Sea-level Rise Could Displace Millions', 20 May 2011, www.irinnews.org/report/92763/vietnam-sea-level-rise-could-displace-millions.

126 Article 23 of UN General Assembly, Universal Declaration of Human Rights.

127 Article 5(e)(i) of UN General Assembly, International Convention on the Elimination of All Forms of Racial Discrimination.

128 Articles 6 and 7 of UN General Assembly, International Covenant on Economic, Social and Cultural Rights.

129 Ibid.

130 Principle 22.b, ECOSOC, Guiding Principles on Internal Displacement.

131 GPC, *Handbook for the Protection of Internally Displaced Persons*, p. 294.

132 See, for example, UNHCR's *Livelihoods and Self-reliance Initiative*, available at www.unhcr.org/pages/4ad2e7d26.html, accessed 13 May 2015.

133 GPC, *Handbook for the Protection of Internally Displaced Persons*, p. 294.

134 Article 8, UN General Assembly, Universal Declaration of Human Rights.

135 Article 2.3, UN General Assembly, International Covenant on Civil and Political Rights.

136 Article 6, UN General Assembly, International Convention on the Elimination of All Forms of Racial Discrimination.

137 Article 14, UN General Assembly, *Convention Against Torture and Other Cruel, Inhuman or Degrading Treatment or Punishment*, 10 December 1984, United Nations, Treaty Series, vol. 1465, p. 85.

138 Article 39, UN General Assembly, Convention on the Rights of the Child.

139 Principle 2, UN General Assembly, Basic Principles and Guidelines on the Right to a Remedy and Reparation for Victims of Gross Violations of International Human Rights Law and Serious Violations of International Humanitarian Law: Resolution/Adopted by the General Assembly, 21 March 2006, A/RES/60/147.

140 IRIN News, '*GLOBAL: Taking on the Land-grabbers*', 26 October 2010, available at: www.irinnews.org/report/90885/global-taking-on-the-land-grabbers.

141 Ibid.

142 Principle 2.1, UN Sub-Commission on the Promotion and Protection of Human Rights, Principles on Housing and Property Restitution for Refugees and Displaced Persons, 28 June 2005, E/CN.4/Sub.2/2005/17.
143 UN General Assembly, Basic Principles and Guidelines on the Right to a Remedy and Reparation for Victims of Gross Violations of International Human Rights Law and Serious Violations of International Humanitarian Law: Resolution/Adopted by the General Assembly.

8 Post-displacement and return

Simon Bagshaw

Every year, millions of people are forcibly displaced from their homes and places of habitual residence as a result of climate-related events such as floods, wind-storms, droughts and other natural hazards. According to the Norwegian Refugee Council's Internal Displacement Monitoring Centre (IDMC), in 2012 alone 32.4 million people were forced to flee their homes by such events, with countries in Asia and West and Central Africa bearing the brunt.[1]

Significantly, 98 per cent of all displacement in 2012 stemmed from climate- and weather-related events. Flood disasters in India and Nigeria accounted for 41 per cent of global displacement in 2012. In India, monsoon floods displaced 6.9 million people, and in Nigeria 6.1 million people were newly displaced. While over the past five years 81 per cent of global displacement has occurred in Asia, in 2012 Africa had a record high for the region of 8.2 million people newly displaced, over four times more than in any of the previous four years.[2] Of course, the scale of displacement is not related simply to climate and weather-related events but is further compounded by such factors as population growth and exposure to natural hazards. As the IDMC observes, migration from rural to urban areas, lack of social housing for poorer families, unplanned growth of informal and unplanned settlements and unimplemented standards for disaster-resistant housing construction puts millions at risk, with the poorest hit the hardest.[3]

As the Peninsula Principles note in the Preamble, while climate displacement can involve both internal and cross-border displacement, most climate displacement will likely occur within State borders. Moreover, as we see in such places as the Philippines, Pakistan and elsewhere, such displacement is often temporary in nature. Those affected will, at some point depending on the context, be able and may or will expect to return to their homes and lands. There are obvious exceptions to this, such as displacement resulting from rising sea levels, where relocation, including across State borders potentially, becomes the only option, and this is where the Peninsula Principles contribution is particularly significant.

Where return is both possible and the desired solution on the part of the displaced population, it must be managed in a way that fully respects their rights. As stipulated in Principle 17 of the Peninsula Principles:

a States should develop a framework for the process of return in the event that displacement is temporary and return to homes, lands or places of habitual residence is possible and agreed to by those affected.

b States should allow climate displaced persons experiencing displacement to voluntarily return to their former homes, lands or places of habitual residence, and should facilitate their effective return in safety and with dignity, in circumstances where such homes, lands or places of habitual residence are habitable and where return does not pose significant risk to life or livelihood.

c States should enable climate displaced persons to decide on whether to return to their homes, lands or places of habitual residence, and provide such persons with complete, objective, up-to-date and accurate information (including on physical, material and legal safety issues) necessary to exercise their right to freedom of movement and to choose their residence.

d States should provide transitional assistance to individuals, households and communities during the process of return until livelihoods and access to services are restored.

This chapter will analyse Peninsula Principle 17 and its essential underpinnings in more detail, as well as some of the practical steps that must be taken to implement them.

In so doing, the chapter will draw on relevant provisions of the Guiding Principles on Internal Displacement. The Guiding Principles were developed by a team of international legal experts working under the direction of the then Representative of the Secretary-General on Internally Displaced Persons, Dr Francis Deng.[4] The Principles were presented to, and taken note of by, the United Nations Commission on Human Rights in 1998.[5] They were subsequently recognized by the General Assembly, including in the Outcome Document of the 2005 World Summit in which they were recognized by the Heads of State and Government as an important international framework for the protection of internally displaced persons.[6]

The Guiding Principles are based on international humanitarian law, human rights law and analogous refugee law. They are intended to serve as an international standard to guide governments, international organizations and all other relevant actors in providing assistance and protection to internally displaced persons. The Principles identify the rights and guarantees relevant to the protection of the internally displaced in all phases of displacement. They provide protection against arbitrary displacement, offer a basis for protection and assistance during displacement, and set out guarantees for safe return, resettlement and reintegration. Although they do not constitute a binding instrument, the Principles reflect and are consistent with international law.

As the Preamble to the Peninsula Principles states, they 'build on and contextualize the United Nations Guiding Principles on Internal Displacement to climate displacement within States'. As such, an understanding of the essential underpinnings of Peninsula Principle 17 requires an understanding of the relevant provisions of the Guiding Principles.

1 The right to voluntary return in safety and dignity

At the core of Peninsula Principle 17, and reflected in paragraphs a and b, is the right to voluntary return in safety and dignity – one of the three so-called 'durable solutions' to displacement. The other two are integration at the location to which people were displaced, or resettlement in another part of the country. As stated in Principle 28(1) of the Guiding Principles:

> Competent authorities have the primary duty and responsibility to establish conditions, as well as provide the means, which allow internally displaced persons *to return voluntarily, in safety and with dignity, to their homes or places of habitual residence, or to resettle voluntarily in another part of the country.* Such authorities shall endeavour to facilitate the reintegration of returned or resettled internally displaced persons [emphasis added].

This Principle stems from the right to liberty of movement and the right to choose one's residence, as embodied in Article 12 of the International Covenant on Civil and Political Rights (ICCPR), which, however, can be limited under certain conditions. Of the durable solutions that exist, voluntary, safe and dignified return of internally displaced persons is often regarded as preferable by displaced persons and authorities alike, at least where this is feasible. As indicated earlier, there are likely to be situations in which climate-displaced persons are simply unable to return to their homes and places of habitual residence, due to sea-level changes for example.

Voluntary return...

While return is often considered the preferred option, rather surprisingly there is no unequivocal rule in existing international human rights treaties that explicitly affirms the right of internally displaced persons to return to their original place of residence, or indeed to move to another safe place of their choice within their own country, although hundreds of international standards and jurisprudence confirm the existence of such rights. The

Guiding Principles on Internal Displacement sought to address this gap through Principle 28(1). In his *Annotations to the Guiding Principles*, Walter Kaelin notes that the right of internally displaced persons to return to their homes and places of origin can be deduced from the above-mentioned ICCPR provision relating to freedom of movement and choice of residence.[7] In addition, the International Labour Organization's (ILO) Convention No. 169 concerning Indigenous and Tribal Peoples states explicitly in Article 16(3) that '[w]henever possible, these peoples shall have the right to return to their traditional lands, as soon as the grounds for relocation cease to exist'. If return is not possible, these peoples shall, pursuant to paragraph 4, 'be provided in all possible cases with lands of quality and legal status at least equal to that of the lands previously occupied by them, suitable to provide for their present needs and future development'. The option of return option is also mentioned, albeit in weaker form, in the 2007 UN Declaration on the Rights of Indigenous Peoples (Article 10).

Support for the existence of a right to return can also be derived from international humanitarian law applicable in situations of armed conflict. Article 49(2) of the Fourth Geneva Convention of 1949, applicable during inter-state armed conflicts, stresses that persons who have been evacuated during an occupation 'shall be transferred back to their homes as soon as hostilities in the area in question have ceased'. Article 85(4)(b) of Additional Protocol I to the Geneva Conventions declares as a grave breach, *inter alia*, unjustifiable delay in the repatriation of civilians when committed wilfully and in violation of the Geneva Conventions and the Protocol. In situations of internal armed conflict, neither common Article 3 nor Protocol II contains rules governing the right of the internally displaced to return to their residences. However, according to the International Committee of the Red Cross, it can be asserted that the right of displaced persons 'to voluntary return in safety to their homes or places of habitual residence as soon as the reasons for their displacement cease to exist' has become part of customary international humanitarian law applicable in both international and non-international armed conflict.[8]

In line with this conclusion, the United Nations Security Council has called on States and the international community to facilitate the return of internally displaced persons. Moreover, on several occasions the Council has explicitly recognized and affirmed the right of such persons to return to their former homes, such as with regard to Bosnia-Herzegovina, Croatia, Georgia and Kosovo,[9] as well as in a thematic resolution on the protection of civilians in armed conflict.[10] The Security Council has also mandated peacekeeping operations under Chapter VII of the United Nations Charter to facilitate the voluntary return of internally displaced persons to their former homes.[11]

Similarly, the General Assembly has reaffirmed the right of all displaced persons to return to their homes or former places of residence in the territories occupied by Israel since 1967.[12] More generally, the former Sub-Commission on Prevention of Discrimination and Protection of Minorities affirmed 'the right of refugees and displaced persons to return, in safety and dignity, to their country of origin and/or within it, to their place of origin or choice'.[13] This right has been reiterated in the Principles on Housing and Property Restitution for Refugees and Displaced Persons.[14]

Among the treaty bodies, the Committee on the Elimination of Racial Discrimination, in General Recommendation XXII(1996) on Article 5 of the Convention on Refugees and Displaced Persons, reaffirmed that all 'refugees and displaced persons have the right freely to return to their homes of origin under conditions of safety'.[15]

At the regional level, the European Court of Human Rights, while explicitly referring to Guiding Principles 18 (right to an adequate standard of living) and 28, has stressed that

> the authorities have the primary duty and responsibility to establish conditions, as well as provide the means, which allow the applicants to return voluntarily, in safety and with dignity, to their homes or places of habitual residence, or to resettle voluntarily in another part of the country.[16]

The right of return for internally displaced persons is embodied in many contemporary peace agreements. Of particular note, Annex 7 of the Dayton Peace Agreement for Bosnia and Herzegovina (DPA) of 14 December 1995 explicitly provided for the right of more than two million refugees and internally displaced persons

> freely to return to their homes of origin. They shall have the right to have restored to them property of which they were deprived in the course of hostilities since 1991 and to be compensated for any property that cannot be restored to them. The early return of refugees and displaced persons is an important objective of the settlement of the conflict in Bosnia and Herzegovina.[17]

... in safety and dignity

While Principle 28(1) articulates a general right to return voluntarily in safety and dignity, Principle 15(d) of the Guiding Principles articulates more fully the safety and dignity aspects of that provision through its articulation of '[t]he right to be protected against forcible return to or resettlement in any place where their life, safety, liberty or health would be at risk'.

As Kaelin observes, prohibiting the return of internally displaced persons to situations of danger can contribute significantly to their physical protection and security, and is corollary to the Principle of voluntariness.[18]

Protection against forcible return to situations of danger is well established in international refugee law through the Principle of *non-refoulement*, and in international human rights law standards relating to torture and the deportation or extradition of aliens.

The Principle of *non-refoulement* in Article 33(1) of the 1951 Geneva Convention relating to the Status of Refugees stipulates:

> No Contracting State shall expel or return *('refouler')* a refugee in any manner whatsoever to the frontiers of territories where his life or freedom would be threatened on account of his race, religion, nationality, membership of a social group or political opinion.

This fundamental Principle of refugee protection is widely regarded as a Principle of customary international law.

In terms of human rights law, Article 3(1) of the Convention against Torture states that '[n]o State Party shall expel, return *("refouler")* or extradite a person to another State where there are substantial grounds for believing that he would be in danger of being subjected to torture.' Article 22(8) of the American Convention on Human Rights states that

> [i]n no case may an alien be deported or returned to a country, regardless of whether or not it is his country of origin, if in that country his right to life or personal freedom is in danger of being violated because of his race, nationality, religion, social status or political opinions.

In refugee law and human rights law, States bear responsibility for violations of the *non-refoulement* Principle and for forcibly returning aliens to situations of danger. The European Court of Human Rights derived the prohibition of return from the prohibition of torture and inhuman treatment in Article 3 of the European Convention on Human Rights, and referred to the 'liability incurred by the extraditing State by reason of its having taken action which has as a direct consequence the exposure of an individual to proscribed ill-treatment'.[19] Similarly, the Human Rights Committee stresses that under Article 7 of the ICCPR, States parties 'must not expose individuals to the danger of torture or cruel, inhuman or degrading treatment or punishment upon return to another country by way of their extradition, expulsion or *refoulement*'.[20]

When this reasoning is applied to the context of internal displacement, 'it is clear that States have a duty to ensure that internally displaced persons are not compelled to return to or be resettled in places where their lives or liberty are at risk.'[21] In the European context, the Council of Europe has now expressly stated that '[i]nternally displaced persons shall not be sent back to areas where they would face a real risk of being subjected to treatment contrary to' the right to life and the prohibition of torture and inhuman or degrading treatment.[22]

2 Implementing voluntary, safe and dignified return – participation, information and assistance

Having restated the Principle of voluntary, safe and dignified return of displaced persons to their original homes or places of habitual residence, Peninsula Principle 17 usefully indicates the sorts of practical measures that are required to give effect to this. Specifically, paragraph c refers to 'enabling climate displaced persons to decide on whether to return to their homes, lands or places of habitual residence through the provision of complete, objective, up-to-date and accurate information (including on physical, material and legal safety issues)'; and paragraph d calls for 'the provision of transitional assistance to individuals, households and communities during the process of return until livelihoods and access to services are restored'.

Participation of the displaced in return planning and management

Paragraph c is essentially about ensuring the full participation of the displaced in the planning and management of their return or resettlement and reintegration. It is a reflection of Guiding Principle 28(2) which states that '[s]pecial efforts should be made to ensure the full participation of internally displaced persons in the planning and management of their return or resettlement and reintegration.' This is not only important for ensuring that such movements are voluntary, but also will greatly facilitate return or resettlement. Guiding Principle 28(2) draws on policy established by the (inter-governmental) Executive Committee (ExCom) of the United Nations High Commissioner for Refugees, in the form of an 'ExCom Conclusion' which stressed the need for refugees to make an informed decision regarding their voluntary return.[23] A subsequent ExCom Conclusion on refugee protection and sexual violence notes that particular safeguards are required to ensure that a refugee woman's decision to repatriate is truly voluntary and not a result of coercion, either direct or circumstantial.[24] As regards internally displaced persons, the Council of Europe has stressed that such 'persons should be properly informed, but also consulted to the extent possible, in respect of any decision affecting their situation' not only prior to and during but also 'after their displacement'.[25]

Access to information

As reflected in Peninsula Principle 17c, a critical component for enabling climate-displaced persons to decide on whether to return to their homes, lands or places of habitual residence and, indeed, for their meaningful participation in the planning and management of their return, is access to complete, objective, up-to-date and accurate information (including on physical, material and legal safety issues). Various initiatives can be pursued to this end.

According to the Inter-Agency Standing Committee's Operational Guidelines on the Protection of Persons in Situations of Natural Disaster,[26] these include: conducting security assessments of sites for return, local integration or settlement elsewhere in the country; establishing comprehensive and accessible public information campaigns as well as grassroots communication strategies on return, local integration and settlement elsewhere in the country; establishing mechanisms such as media reports, databases, information centres, etc. to provide internally displaced persons with information on the conditions at the place of their former homes or locations identified for settlement elsewhere in the country, and organization of 'go and see' visits.

Relevant measures would also include identifying persons with special needs and including them in the planning and management of return, local integration or settlement elsewhere in the country, including through outreach activities and focus group meetings where appropriate; publishing and widely disseminating zoning and rebuilding plans and holding planning commission meetings that are open to the general public; monitoring and identifying instances of discrimination, in particular of persons with special needs, in providing access to durable solutions including adequate housing, basic services and livelihoods; and removal of legal and administrative obstacles that hinder local integration or settlement elsewhere in the country.

Supporting sustained return

Of course, it is not sufficient simply to return people to their homes and lands. If return is to be sustainable then it must be supported, at least in the short to medium term. As stipulated in Peninsula Principle 17d, transitional assistance should be provided to individuals, households and communities during the process of return until livelihoods and access to services are restored. Such assistance may be varied in nature and will often be dictated by the context. Broadly speaking, it would encompass access to an adequate supply of safe food and nutrition; to safe and potable water and adequate sanitation; to adequate shelter; and to health and education services.[27]

In many, if not most, situations, international humanitarian and development organizations will play a key role in the provision of such assistance. While this is not expressly recognized in Peninsula Principle 17, Principle 30 of the Guiding Principles stipulates that '[a]ll authorities concerned shall grant and facilitate for international humanitarian organizations and other appropriate actors, in the exercise of their respective mandates, rapid and unimpeded access to internally displaced persons to assist in their return or resettlement and reintegration.'

Property restitution

Though not expressly recognized in the Peninsula Principles, a critical component in many contexts for the sustainability of return is the restitution of property that displaced persons left behind. This is recognized in the Guiding Principles which, in Principle 29(2), stipulate that:

> Competent authorities have the duty and responsibility to assist returned and/or resettled internally displaced persons to recover, to the extent possible, their property and possessions which they left behind or were dispossessed of upon their displacement. When recovery of such property and possessions is not possible, competent authorities shall provide or assist these persons in obtaining appropriate compensation or another form of just reparation.

Whether as a result of climate-related disasters or armed conflict, internally displaced persons regularly lose access to their property when displaced. When they return to their former habitual residence they may find their properties destroyed, confiscated, expropriated or occupied by other people, raising questions of whether they have a right to restitution for the property or to compensation for its loss.

As Kaelin observes, there is a certain trend in general human rights instruments, along with the progressive development of international law, to answer these questions in the affirmative[28] insofar as they guarantee the right to property[29] or, in some cases, the rights to be free from arbitrary interference with one's home[30] and to adequate housing.[31]

Also relevant in this context is the International Court of Justice's advisory opinion of 9 July 2004 on the Legal Consequences of the Construction of a Wall in the Occupied Palestinian Territory, in which it referred to the fundamental Principle that breaches of international law entail a duty to provide reparation that

> must, as far as possible, wipe out all the consequences of the illegal act and reestablish the situation which would, in all probability, have existed if that act had not been committed. Restitution in kind, or, if this is not possible, payment of a sum corresponding to the value which a restitution in kind would bear; the award, if need be, of damages for loss sustained which would not be covered by restitution in kind or payment in place of it – such are the Principles which should serve to determine the amount of compensation due for an act contrary to international law.[32]

The Court concluded that these Principles apply to reparation in the form of restitution of or compensation for 'the requisition and destruction of homes, businesses and agricultural holdings' owned by natural or legal persons that was a consequence of the construction, in violation of international human rights and humanitarian law, of the wall in the Occupied Palestinian Territory.[33]

As regards the issue of restitution, the Security Council has affirmed the importance of internally displaced persons being able to return 'to their homes and property', as well as the fact that 'individual property rights have not been affected by the fact that owners had to flee during the conflict and that the residency rights and the identity of those owners will be respected'.[34] This resolution reflects the United Nations Secretary-General's call to take

> [r]estorative actions, such as the inclusion of the right to return and restitution of housing, land or property in all future peace agreements and all relevant Council resolutions, and the inclusion of housing, land and property issues as an integral part of future peacekeeping and other relevant missions, with provisions for dedicated, expert capacity to address these issues.[35]

Principle 2.1 of the above-mentioned Pinheiro Principles provide that refugees and internally displaced persons

> have the right to have restored to them any housing, land and/or property of which they were arbitrarily or unlawfully deprived, or to be compensated for any housing, land and/or property that is factually impossible to restore as determined by an independent, impartial tribunal.

Pursuant to Principle 16.1, the Principles extend these rights to 'tenants, social occupancy rights holders and other legitimate occupants or users of housing, land and property' and stress that such claimants should, 'to the maximum extent possible', be 'able to return to and re-possess and use their housing, land and property in a similar manner to those possessing formal ownership rights'.

The Pinheiro Principles also address the problem of 'secondary occupants', i.e. persons (often refugees or internally displaced persons themselves) who were allowed to use property left behind by the displaced. They take the approach that the rights of the original owners are stronger than those of such occupants. However, they call upon States to ensure 'that secondary occupants are protected against arbitrary or unlawful forced eviction' and that evictions which are unavoidable to return property to the original owners 'are carried out in a manner which is compatible with international human rights law and standards', i.e. with 'an opportunity for genuine consultation, adequate and reasonable notice, and the provision of legal remedies, including opportunities for legal redress' (Principle 17.1).

Restitution versus compensation

As stipulated in Guiding Principle 29(2), when recovery of property and possessions is not possible, competent authorities shall provide or assist these persons in obtaining appropriate compensation or another form of just reparation. Although compensation may in certain circumstances be a favourable alternative to restitution it is by no means a panacea. For example, in the case of cash compensation there will likely be difficulties in determining what constitutes an equitable sum. In addition, the ways in which money is used or distributed between family members raise the risks of disputes and inequities within the household that may not be so acute in the case of restoration of land and property rights. In particular, and as will be discussed further below, issues related to gender relations and age differences amongst the household may be problematic. Moreover, the payment of cash compensation may only serve to compound the situation of those displaced. Throwing money at displaced persons whose livelihoods are dependent on access to land, such as farmers and pastoralists, will not necessarily solve their problems in the same way as would allocation of equivalent land elsewhere in the region or country. It is with this in mind that Principle 29(2) stipulates that when restitution is not possible the authorities will assist the displaced in obtaining '*appropriate* compensation or another form of *just* reparation' (emphasis added).

The Pinheiro Principles also address this issue and in doing so give clear priority to restitution when they state

> that the remedy of compensation is only used when the remedy of restitution is not factually possible or when the injured party knowingly and voluntarily accepts compensation in lieu of restitution, or when the terms of a negotiated peace settlement provide for a combination of restitution and compensation.[36]

Restitution, compensation and equal access for women

Another important aspect to the right to restitution which should be considered is the need to ensure equal access of internally displaced women to restitution or compensation mechanisms. First, internally displaced populations are invariably composed of greater numbers of women and children than men. Second, in parts of Africa and Asia women are often discriminated against in terms of inheritance rights. That is to say, when a man dies it is common for his relatives to take possession of all property belonging to the couple, including land and the family house, often leaving the widow deprived of an economic base and having to support herself and her children in extreme poverty. The existence of such practices may have repercussions for returning internally displaced women – whose husbands have died either prior to or during displacement – both in terms of establishing their legal rights to the property in the first place, as well as in gaining access to restitution procedures. In this regard, it would seem pertinent to read Guiding Principle 29 in conjunction with Guiding Principle 4 according to which:

1   These Principles shall be applied without discrimination of any kind, such as race, colour, sex, language, religion or belief, political or other opinion, national or social origin, legal or social status, age, disability, property, birth, or any other similar criteria.

2   Certain internally displaced persons, such as children, especially unaccompanied minors, expectant mothers, mothers with young children, *female heads of household*, persons with disabilities and elderly persons, shall be entitled to protection and assistance required by their condition and to treatment which takes into account their special needs [emphasis added].

The Pinheiro Principles are more specific on this point, including in terms of detailing specific measures that should be taken to ensure equal access for women. Pursuant to Principle 4:

4.1   States shall ensure the equal right of men and women, and the equal right of boys and girls, to housing, land and property restitution. States shall ensure the equal right of men and women, and the equal right of boys and girls, *inter alia*, to voluntary return in safety and dignity, legal security of tenure, property ownership, equal access to inheritance, as well as the use, control of and access to housing, land and property.

4.2   States should ensure that housing, land and property restitution programmes, policies and practices recognise the joint ownership rights of both male and female heads of the household as an explicit component of the restitution process, and that restitution programmes, policies and practices reflect a gender-sensitive approach.

4.3   States shall ensure that housing, land and property restitution programmes, policies and practices do not disadvantage women and girls. States should adopt positive measures to ensure gender equality in this regard.[37]

Peninsula Principle 7 provides general provisions to this effect. In particular, paragraphs d and e stipulate that:

d   States should ensure the right of all individuals, households and communities to adequate, timely and effective participation in all stages of policy development and implementation of these Peninsula Principles, ensuring in particular such participation by indigenous peoples, women, the elderly, minorities, persons with disabilities, children, those living in poverty, and marginalized groups and people.

e   All relevant legislation must be fully consistent with human rights laws and must in particular explicitly protect the rights of indigenous peoples, women, the elderly, minorities, persons with disabilities, children, those living in poverty, and marginalized groups and people.

Conclusion

Where return is both possible and the desired solution on the part of the displaced population, it must be managed in a way that fully respects their rights. The Peninsula Principles provide important contextual guidance in this respect which, when read in conjunction with the Guiding Principles on Internal Displacement and other relevant standards, provide a very comprehensive framework of protection and assistance.

Notes

1 Internal Displacement Monitoring Centre, *Global Estimates 2012 – People Displaced by Disasters* (Geneva: IDMC, 2013).
2 Ibid.
3 Ibid., p. 12.
4 See Report of the Representative of the Secretary-General on Internally Displaced Persons, E/CN.4/1998/53/Add.2 (1998). For an overview of the process leading to the development of the Principles, see Simon Bagshaw, *Developing a Normative Framework for the Protection of Internally Displaced Persons* (Ardsley, NY: Transnational Publishers, 2004).
5 Commission on Human Rights resolution 1998/50.
6 See General Assembly resolution 60/1 (2005).
7 Walter Kaelin, *Annotations to the Guiding Principles on Internal Displacement,* (Washington DC: American Society of International Law, 2008), p. 127.
8 Rule 132. See Jean-Marie Henckaerts and Louise Doswald-Beck, *Customary International Humanitarian Law. Volume I: Rules* (Cambridge: Cambridge University Press, 2005).
9 See, respectively, UN Security Council resolutions S/RES/820 (1993), para. 7; S/RES/1009 (1995), para. 2; S/RES/876 (1993) and S/RES/1781 (2007), para. 15; and S/RES/1244 (1999).
10 See S/RES/1674 (2006).
11 See, for example, S/RES/1778 (2007), para. 1 (Chad); and S/RES/1756 (2007), para. 3(b) (Democratic Republic of Congo).
12 See A/RES/51/126.
13 Sub-Commission Resolution 94/24, E/CN.4/Sub.2/1994/56.
14 See Housing and property restitution in the context of the return of refugees and internally displaced persons. Final report of the Special Rapporteur, Paulo Sérgio, Pinheiro Principles on Housing and Property Restitution for Refugees and Displaced Persons, E/CN.4/Sub.2/2005/17 (2005). Endorsed by the United Nations Sub-Commission on the Promotion and Protection of Human Rights on 11 August 2005. The Principles resulted from a seven-year process which initially began with adoption of Sub-Commission resolution 1998/26 on housing and property restitution in the context of the return of refugees and internally displaced persons in 1998. This was followed from 2002 to 2005 by

a study and proposed Principles by the Sub-Commission Special Rapporteur on Housing and Property Restitution, Paulo Sérgio Pinheiro. See further: OCHA *et al.*, *Housing and Property Restitution for Displaced Persons and Refugees – Implementing the Pinheiro Principles* (March 2007).

15 CERD, General Recommendation XXII – Refugees and Displaced Persons (1996), para. 2(a).
16 European Court of Human Rights, *Doğan* v. *Turkey*, Judgment of 29 June 2004, para.154.
17 Article I(1) of Annex 7.
18 Kaelin, *Annotations*, p. 69.
19 European Court of Human Rights, Cruz Varas Case, Judgment of 20 March 1991, Series A, No. 201, para.69.
20 Human Rights Committee, General Comment No. 20, Article 7: Prohibition of torture or cruel, inhuman or degrading treatment or punishment.
21 Kaelin, *Annotations*, p. 70.
22 Council of Europe, Committee of Ministers, Recommendation (2006) 6, para. 5.
23 ExCom Conclusion No. 18 (XXXI/1980) on Voluntary Repatriation.
24 ExCom Conclusion No. 73 (XLIV/1993) on Refugee Protection and Sexual Violence. UNHCR's *Handbook for the Protection of Women and Girls* (Geneva: UNHCR, 2008) contains a broad range of actions that can be taken to this end. These include: promoting respect for women's and girls' equal rights to make a free and informed choice to return voluntarily and to equal access to housing, land, property and inheritance so as to enable female refugees to return, including by ensuring that information relevant to their concerns is transmitted and incorporated into information provided to the community deciding to return; working with women and girls and the whole community to develop appropriate voluntary return methodologies that take into account the concerns of specific groups, particularly female-headed households, women and girls with disabilities, older women with no family, and women and girls who have been subject to rape and other forms of sexual and gender-based violence by persons in their area of return. In addition, efforts should be made to involve women, men, girls and boys in the design of reconstruction and assistance pro-grammes and community-based economic projects in return areas and support women to ensure that they benefit equitably from the projects established. See further, UNHCR, *Handbook for the Protection of Women and Girls*, pp. 154–156.
25 Council of Europe, Committee of Ministers Recommendation 6 (2006), para. 11.
26 IASC, *Operational Guidelines on the Protection of Persons in Situations of Natural Disasters* (Washington DC: IASC and Brookings-Bern Project on Internal Displacement, 2011), pp. 47–48.
27 Ibid., pp. 29–38.
28 Kaelin, *Annotations*, p. 132.
29 See, for example, Article 17 UDHR, Article 21 ACHR, Article 14 AfCHPR, Article 31 ArCHR, and Article 1 Protocol No. 1 to the ECHR.
30 Article 12 UDHR, Article 17 CCPR, Article 11 ACHR, Article 21 ArCHR, Article 8 ECHR.
31 Article 25 UDHR, Article 11 CESCR, Article 27 CRC, Article 26 ACHR in conjunction with Article 31 (k) of the 1970 Buenos Aires Protocol to the Charter of the Organization of American States, Article 38 ArCHR, and Article 31 (1) of the 1996 revised European Social Charter; see also Article 5 (e) (iii) CERD and Article 14 (2) (h) CEDAW regarding non-discrimination in the area of housing.

32 International Court of Justice, Legal Consequences of the Construction of a Wall in the Occupied Palestinian Territory, Advisory Opinion, 9 July 2004, para.152.
33 Ibid.
34 S/RES/1781 (2007), operative para.15.
35 Report on the Protection of Civilians in Armed Conflict, S/2007/643, para. 59.
36 Principle 21.1.
37 In practical terms, this means implementing a range of possible actions, including the design of special programmes aimed at assisting women and girls to make restitution claims, gender-sensitivity training for officials entrusted with working on restitution matters, providing special outreach about restitution issues to women's organizations or women's networks, and/or providing special resources to households headed by single women so that they are also able to avail themselves of their housing and property restitution rights. Efforts should also be made by international and domestic actors to monitor women's housing and property restitution rights and should include coverage of any sexual or gender-based violence carried out by State or non-State actors, particularly when this violates the rights of women to return to their homes 'in safety and dignity'. See further, OCHA *et al.*, *Housing and Property Restitution*, pp. 36–37.

9 Implementation

Khaled Hassine

Introduction

Protecting climate-displaced persons demands a set of complex legal, institutional and ultimately managerial and social interventions, which take into account the uniqueness of each country while building on the lessons learned from similar scenarios. The Peninsula Principles constitute a critical set of Principles that are designed to be adapted to each country's contextual factors, its own set of key actors, institutions and vulnerable populations.

This chapter argues that the global consultative process and the bottom-up approach that led to the formulation and adoption of the Peninsula Principles do not only ensure the appropriateness of this framework in dealing with future and present situations of climate displacement in that the Peninsula Principles address the needs of those affected; they also ensure acceptability and ultimately implementation of the provisions. When considering the very nature of the Principles, which are derived from existing laws and standards for the purpose of addressing situations of climate displacement, implementation becomes not only a possible option, but rather an imperative that is further corroborated by the flagrant need for international guidance on how to address the effects of climate change. Some of the legal and practical challenges associated with climate displacement are addressed by the current international framework. Others, however, are not explicitly or are only inadequately dealt with and the Peninsula Principles therefore fill a gap spanning the entire displacement continuum and moving from the reactive to the preventive sphere. This chapter further addresses the implementation *modus operandi* as well as the main actors and concludes by providing an overview of some of the specific elements of the implementation portfolio.

Global consultative process

One distinct feature of the Peninsula Principles is the process through which they were articulated. The Principles' genesis is indeed remarkable as they are the result of an international legislative process that was not steered by lawmakers in a multilateral framework. They emerged from a bottom-up process involving climate-affected people and communities from numerous countries, growing out of the necessity to address climate displacement in a comprehensive manner. This constitutes the foundation to successfully address the needs of climate-displaced persons. It is the affected people themselves who felt that there was a pressing need to develop a normative, institutional and implementation framework.

The Climate Displacement Law Project initiative, under the umbrella of Displacement Solutions,[1] provided the required platform for consultations and continuous cooperative work with UN officials, governments, climate change experts, legal scholars, academics, non-governmental and community-based organizations around the world. This international non-governmental organization responded to grassroots requests for guidance and solutions, and helped to facilitate and steer a process geared towards addressing the pivotal questions of climate displacement that concern people everywhere.

The consultative process spanned several years and included preparatory work and developments on the ground, including capacity-building, mobilization and advocacy, and then seamlessly integrated them into a global consultative process involving researchers and experts as well as governments, international organizations and United Nations agencies. This process leading up to the adoption of the Principles is also distinct in that it provided room for public comments on preliminary drafts. It ultimately culminated in the endorsement of the Principles by the drafting group composed of representatives from Alaska, Australia, Bangladesh, Egypt, Germany, the Netherlands, New Zealand, Switzerland, Tunisia, the United Kingdom and the United States of America, who combined their expertise in international law, migration and environmental change. In the light of this particular legislative history, the Peninsula Principles reflect the needs of those concerned and are by their very nature rights-based and people-centred.

Global *minimum proprium*

The Peninsula Principles are derived from existing international legal standards for the purpose of addressing situations of climate displacement. One of the Principles' achievements is that they succeeded in bringing together the norms that are scattered in different legal fields, including provisions applicable to internally displaced persons, humanitarian law (particularly on issues pertaining to refugees) and human rights law, as well as housing, land and property rights (HLP) protection. The different legal regimes often operate in silos and are not easily integrated in a cohesive manner, particularly when providing the foundation for the operations of distinct actors such as humanitarian workers drawing on refugee law, and human rights officers who base themselves on the human rights body of law and follow a human rights-based approach in their daily work.[2]

With the Peninsula Principles, the various standards that are also applicable to situations of climate displacement are now consolidated into one normative document as a global *minimum proprium*. They are designed to provide guidance to States, displaced communities, international organizations, non-governmental and community-based organizations and other relevant stakeholders, including private corporations in their role as service providers,[3] on how to best address the complex legal, technical and practical issues surrounding climate displacement within States. They can be referred to as a crucial source of international standards supporting the rights of disaster-affected populations to return to and recover housing, land and property rights.

The Principles are deeply grounded in existing international public and human rights law. They constitute a restatement of the law, which they seek to tailor to the specific needs of climate displacement, addressing existing gaps and formulating policy and action guidance. Beyond these minimum standards, reflecting well-established good practice to address the rights of climate-displaced persons, the Principles formulate concrete policy and management measures in order to respond to situations of climate displacement in a human rights-compatible manner.

It seemed more appropriate to prepare a set of Principles that would immediately serve as a standard, rather than to engage in potentially protracted negotiations on a new convention on climate-displaced persons, which would then require a certain number of ratifications to enter into force, thus further delaying the promulgation of a comprehensive framework that is so desperately needed. As a matter of fact, as the Principles restate existing norms, they are of significant legal value in spite of their non-binding character. They gain authoritative value as an alternative international legislation through the mere fact that they do not create new law, but constitute a reaffirmation and clarification of the existing framework. As such, the Principles, as part of the body of norms within legal systems that demand realization of the purported objective to the greatest extent possible,[4] seek to empower States in dealing with climate displacement as they stand for a commitment to the results in an optimized manner.

By contrast, a new convention is something that may take years eventually to materialize, and moreover seems to be a challenging option at the moment from the political perspective.[5] As a matter of fact, a number of key donors actively and financially supported the conceptual approach of the Peninsula Principles, which is respectful of State sovereignty. This notable aspect of the *travaux préparatoires* indicates that governments, for the time being, espouse the elaboration of Principles on climate displacement with a focus on intra-State movement rather than the development of a new convention, and also suggests a lack of political will with regard to the latter.

Sticking to the existing framework is of critical importance for the Peninsula Principles' broad acceptance, political endorsement and ultimately, their effective implementation. The interplay with and respect for the Nansen Initiative[6] governed the drafters' choice to limit the applicability of the Principles to intra-State displacement. This effectively eliminates a number of controversial issues that could block political consensus on a climate displacement framework, notably the question of cross-border migratory movements as a result of climate change, whereby the complex issue of causality comes into play. Focusing the scope of the Principles on intra-State displacement is moreover a legitimate prioritization and will tackle the core of the problem as by now it is widely accepted that most climate displacement will occur within States. Cross-border movement is being dealt with by the Nansen Initiative, which focuses on the needs of persons displaced across borders and aims at developing a protection agenda in this regard. It is this initiative that seeks to address the question of protection of people who are forced to leave their countries for reasons such as the disappearance of their own State, as in the case of the 'sinking islands'. The complexity of the issue is illustrated by a recent judgment of the New Zealand High Court concerning an i-Kiribati national who claimed asylum as a climate refugee. He argued that global warming was a form of persecution and those displaced across international borders due to the effects of climate change should be recognized as refugees under the 1951 Refugee Convention. In particular, the Court found that 'it is not for the High Court of New Zealand to alter the scope of the Refugee Convention in that regard. Rather that is the task, if they so choose, of the legislatures of sovereign states.'[7] Complementary to the Nansen Initiative, the Peninsula Principles focus on the displacement due to the effects of climate change *within* the borders of a State.

These two aspects, the global participatory nature of an all-inclusive consultative process translating into a people-centred approach, and the fact that the Principles constitute the global *minimum proprium*, form the basis for an immediate implementation imperative and direct applicability of the Peninsula Principles in tandem with the ongoing global governmental and advocacy process.

Filling the gap

Climate adaptation, which includes among other strategies migration, displacement and planned relocation, raises a number of legal and practical questions. Some of these challenges are addressed by the current international framework, while others are not explicitly or only inadequately dealt with.[8]

The Guiding Principles on Internal Displacement is the most important normative document in relation to internal displacement, a category which also includes persons displaced as a result of climate change. As a matter of fact, the Peninsula Principles build on the Guiding Principles and con-textualize them to climate displacement situations within States. The Guiding Principles are not a legally binding document, but – like the Pen-insula Principles – they are a restatement of existing norms firmly grounded in international law[9] and seek to provide internally displaced persons with the full range of human rights guarantees, including protection against arbitrary or forced displacement.

The Guiding Principles are also applicable to relocations in the context of climate change, but only address the broader issue of internal displace-ment, and do not, beyond the general Principle, provide practical guid-ance in terms of the handling of climate displacement situations. When it comes to operational guidance, this can be found in the Inter Agency Standing Committee (IASC) Operational Guidelines for Protection of Persons in Situations of Natural Disasters and the IASC Framework for Durable Solutions, which have concretized the Guiding Principles and constitute important derivates.[10] These also inspired the Peninsula Prin-ciples. Moreover, the World Bank's Operational Policy 4.12 on Involun-tary Population Resettlement provides detailed guidance in the context of relocations, though this policy seeks to provide internal guidance and operating procedures for World Bank staff, and is not an affirmation of rights. They were moreover designed as a response to displacement situations resulting from infrastructure projects and are not tailored to climate displacement. Other important tools are the safeguards developed by the World Bank on involuntary resettlement, though many have criticized deficiencies regarding the protection of the right to adequate housing.[11]

Moreover, and this is one of the major lacunae of the Guiding Principles as a framework applicable to climate displacement, they do not apply to internal displacement as a result of slow-onset disasters, but are limited to internally displaced persons (IDPs) uprooted by sudden-onset disasters. This constitutes a significant protection gap, since climate change is par-ticularly associated with this type of slow-onset environmental degrada-tion, and disasters, though their occurrence may increase, are projected to remain episodic. By contrast, the scope of the Peninsula Principles is wide and they are inclusive in nature. They apply to all persons displaced within a State due to the effects of climate change, including sudden-onset as well as slow-onset disaster scenarios, and are as such distinct from the Guiding Principles. The issue of path-dependency[12] between displacement and climate change in the case of slow-onset disasters has become obsolete with the Peninsula Principles, in particular through the formula that the effects of climate change may occur either alone or in combination with other factors and most importantly through the incorporation into the Principles of a preventive approach, specifically tailored to climate change.[13] In giving effect to the latter, it is for States to determine locations that are threatened by climate change, including the effects of slow-onset disasters. In a cross-border displacement scenario this may still be a contentious matter, but in the case of displacement within States, which circumscribes the application of the Peninsular Principles, this approach will for instance allow States to identify those individuals who qualify as climate-displaced persons as they move from a specified risk area.

Another set of guidelines that apply to post-disaster situations and climate displacement are the Pinheiro Principles.[14] These Principles apply to *all* displaced persons, irrespective of the causes of displacement and their location, i.e. including those displaced due to the effects of climate change. However, their focus lies on HLP rights protection by providing *all* refugees and displaced persons with a distinct and genuine individual right to claim back their homes, land or property.[15]

Also significant at the regional level is the Kampala Convention,[16] which provides for the protection and assistance of persons displaced as a result of natural disasters and climate change. This agreement, though regional in nature, has potential significance for other regions and constitutes an important precedent.

In 2007, the IASC Early Recovery Cluster had already identified the need for specific guidelines on post-disaster land responses. While there are many similarities between post-disaster and post-conflict land issues, there are a number of important differences and distinctive imperatives of post-disaster land programming and it is essential to adopt a perspective that is adapted to the particularities of natural disaster settings.[17]

Beyond the need for a globally applicable normative framework such as the Peninsula Principles, there has been no coordinated response by States to address climate displacement and there is a pressing need for guidance and expertise in this regard. The Republic of Georgia is a good example of a country that is faced with the challenge of what is called 'eco-migration' and illustrates the needs for the Peninsula Principles. There have been various attempts to respond to natural disasters in Georgia's mountain regions, starting with the Soviet administration, by resettling the affected population, but there has not been until now any comprehensive policy or strategy on how to effectively deal with this 'eco-migration'.[18] As a result of the lack of a comprehensive policy or long-term strategy in the Georgian context there is no systematic assessment procedure of present or future needs and policy therefore remains reactive. Decision-making is often discretionary without any form of oversight and accountability framework. Resource allocation does not match the magnitude of the problem and therefore remains inadequate. The lack of a comprehensive action plan also translates into a narrow focus on housing and land purchases for resettlement, without, however, providing any resettlement assistance, which is central to the successful integration of the displaced into the host communities.[19]

Other recent experiences, be it the 2010 earthquake in Haiti, the 2004 tsunami in the Indian Ocean or Hurricane Katrina which struck New

Orleans in 2005, also demonstrate that a comprehensive approach to post-disaster and climate displacement is missing and that this is detrimental to the effective enjoyment of their rights by the affected populations.

In the post-tsunami period, for instance, some governments have misused adaptation strategies.[20] With the argument of an imminent protection need, authorities in some States adopted policies which established coastal buffer zones; these prevented low-income affected persons from returning to their areas of origin and forced communities in coastal areas to relocate, thereby disrupting livelihoods, and generating and exacerbating social tensions. Fishing communities were particularly affected by these measures which permanently prevented them from returning to their original homes. Moreover, in some instances such buffer zones were transferred to higher-income residential, commercial or industrial use, as for instance the tourism industry expanded its operations into the newly vacated land.[21]

Delay in return in Haiti as a result of entrenched patterns of discrimination and neglect and the comparatively more attractive conditions in the camps demonstrated the need for a comprehensive approach focusing on communities rather than individuals. This is another case underlining the need for the Peninsula Principles.[22]

New Orleans exemplified another facet of the risks in post-disaster settings. The privatization of housing, education and health care and the process of 'green-spacing' by urban planners, transforming residential areas at risk into non-residential parks, resulted in the forced displacement of entire communities.[23]

The need for international guidance on how to address the effects of climate change is further exemplified by the post-Katrina jurisprudence or rather the lack of a decision on the merits in a case regarding liability for the disaster. Victims of Hurricane Katrina contended that several energy companies were responsible for the increased destructiveness of the hurricane because of their greenhouse gas emissions. The victims sought compensation for loss of private property and use of public property, alleging nuisance, negligence, trespass, unjust enrichment, civil conspiracy and fraudulent misrepresentation. In a landmark decision a US Court of Appeals ruled that the victims had grounds to bring their public and private nuisance claims, as well as trespass and negligence claims, hence that they could seek relief for property damages resulting from Hurricane Katrina, and that none of these claims present non-justiciable political questions. When, however, defendants asked for rehearing *en banc*, which was granted, eight judges recused themselves. As the court no longer had a quorum, the case could not be reviewed. And because the panel decision was vacated, the trial court dismissal was valid. When plaintiffs filed their claim again, the trial court dismissed it for the second time, on grounds of *res judicata* as well as the statute of limitations, the political question doctrine, pre-emption, proximate cause and standing.[24]

Displacement and protection continuum

Another essential feature of the Peninsula Principles is that they span the entire displacement continuum, i.e. they provide guidance at all stages of climate displacement – prior, during and after displacement – and are not therefore merely reactive. The Principles are not only about existing displacement settings, but also about avoiding future displacement scenarios and putting in place a legal framework to address future situations. The approach followed is rather to provide pre-emptive assistance to those who may be displaced by the effects of climate change, in addition to the effective remedial assistance to those already displaced. Legal protection extends to both existing and future climate-displaced persons.

The Peninsula Principles represent a radical shift in addressing climate displacement as they advocate a pre-emptive approach in dealing with this phenomenon under the absolute condition that land solutions in the context of a rights-based and participatory relocation process should be found for those who cannot stay in or return to their homes. A pre-emptive approach is fundamental to the identification of suitable solutions for future relocation in order to avoid resettlement sites becoming subject to speculation, which is a frequent feature of relocation endeavours.

This is in line with States' obligations regarding disaster preparedness, early warning and judicial response as outlined by the Budayeva jurisprudence.[25] This Russian case concerns the destruction of residents' homes by a mudslide. According to the applicants, the authorities knew about the risk of mudslide, and failed to take necessary preventive action (e.g. reinforcing a dam wall and warning the residents of the imminent risk). The court looked at a number of aspects, including the authorities' failure to set up a warning system and the adequacy/inadequacy of a protective infrastructure, as well as the judicial response to the disaster. The court concluded that there had been no justification for the authorities' failure to implement land planning and emergency relief policies in the hazardous area giving the foreseeable risk to the lives of its residents. Moreover, it found that the serious administrative flaws, which had prevented the implementation of those policies, had caused the death of members of the applicants' family. The authorities had therefore failed in their duty, as a form of due diligence obligation, to establish a legislative and administrative framework with which to provide effective deterrence against a threat to the right to life.[26]

Planning is a central and very distinct aspect of the Peninsula Principles. One of the main challenges today for relocation programming is the lack of lead-time for planning and consultation. As opposed to development-induced resettlement, disaster-related relocation is currently often reactive in nature.[27] Particularly in the case of slow-onset disasters, it will, however, be possible to plan and prepare for future displacement scenarios and to engage in activities to reduce vulnerability in the areas concerned. Spatial planning mechanisms, for instance, that restrict reconstruction in inappropriate or unsafe locations, and relocation of infrastructure and government facilities to safe areas can minimize the impact of future disasters. Other important activities include anticipating climate displacement through systematic data collection, observation and monitoring; assessing relocation needs; as well as the early identification of suitable and habitable land and the planning and development of relocation sites including human settlements. It needs to be borne in mind that climate displacement preparedness and risk management require a high degree of capacity and coordination among local institutions involved in land planning and relocation, and may thus require international assistance.

One of the challenges in this regard is the way in which international assistance operates. The issue of climate displacement is governed by a number of different institutions across several policy fields, including the humanitarian, human rights and development fields. This potentially results in overlaps, and also increases the risk of gaps and inconsistency in responding to climate displacement. Different institutions with authority over the same issue may implement parallel activities. There is a need for enhanced coordination and a holistic approach to implementation, and a necessity to clearly define roles and responsibilities within transparent institutional arrangements, so as to effectively capitalize on the strengths of each stakeholder.[28]

Appliance *modus operandi*

It is essential to apply the Peninsula Principles in practice through various programmes, since not all the challenges in relation to climate displacement, particularly the practical ones, can be overcome by legal means. It is rather the actual implementation that will ensure protection of climate-displaced persons.

A distinct added value of the Principles is that they constitute a baseline for further concrete discussions. They provide those elements that are indispensable when addressing climate displacement and that need to be translated into concrete action. Obviously, the Principles do not prescribe in a rigid manner the way in which they ought to be concretized. They are deliberately formulated in a manner which recognizes the similarities and, at the same time, the specificities of each displacement scenario and therefore leaves the decision-makers with enough room to adapt the provisions to the specific settings. By virtue of being *prima facie*, the ways in which the Principles are applied must be determined and optimized anew in every new setting, and in light of all known relevant circumstances.[29] Hence they provide the basis for the elaboration of detailed and tailored national action plans in order to address present and future climate displacement, and to design implementation frameworks and procedures.

The Principles can also serve as an instrument to evaluate policy coherence, meaning that national laws and policies can be assessed against this minimum standard in order to identify shortcomings and gaps. This contributes to national legal preparedness.

Support from key donor countries, as received in the course of the consultative and the drafting process of the Principles, is central in order to build broad-based consensus around the Principles and is the driving force behind actual implementation. Actual implementation does not require adoption at the international level through any particular forum but is concerned with effective application of the Principles' provisions, i.e. their traceability in legislative and practical action at the domestic and international levels. The adoption of national laws and policies drawing from the Peninsula Principles reinforces the Principles' authoritative value and has a multiplying effect on further implementation. References in jurisprudence, legislation and more generally the body of human rights law, such as reports and recommendations of Special Procedures, also support the Principles' dissemination. United Nations human rights mechanisms, including the Human Rights Council, Special Rapporteurs and Independent Experts, the Universal Periodic Review and Treaty Monitoring Bodies, should include climate displacement concerns when examining State reports and conducting country visits, and identify main challenges as well as best practices at the national levels. Their analysis is an important source for further action.[30]

Furthermore, tailored advocacy for the Peninsula Principles, including through a dissemination plan to reach governments and civil society, is crucial to raise awareness of the Principles and facilitate their implementation and sustained usage. One of the challenges in this regard is the need to strengthen the capacity of national authorities as well as non-governmental and community-based organizations to effectively and correctly implement and apply the Principles.

Mapping climate displacement actors

The primary addressee of the Peninsula Principles is the State. It is also the State which retains the primary responsibility for their implementation. Non-State actors who are effectively exercising control over a territory to the extent that the rights of the climate-displaced are affected also carry responsibility for upholding the obligations under the Principles. Yet, in cases of non-existence of a State entity or in cases of a non-identifiable State-actor, and where the State does not have sufficient capacity, this obligation extends to the international community and international agencies.

In order to meet the climate displacement challenge, it is necessary to engage leadership and to work in partnership with all sectors of society to identify and highlight solutions. This has recently been acknowledged by the Secretary-General of the United Nations, who devised a climate strategy to promote climate action among stakeholders, including governments, civil society, business and finance, and is providing support to strategic, policy and political aspects in this regard.[31] In particular it is crucial to fully engage the private sector regarding the construction of resilient infrastructure, sustainable development of urban areas, energy safety, and the protection of critical resources.[32]

The Principles acknowledge the key role played by other actors in the implementation of these provisions, and require the State to cooperate with inter-governmental organizations, non-governmental and community-based organizations and civil society, as well as with practitioners. In doing so, States should develop a comprehensive assistance framework, detailing the intended beneficiary communities and specific destinations for relocation, as well as specific financial information, which would then need to be submitted to interested international donors for assistance and action. The elaboration and subsequently the implementation of such a framework have to be carried out in consultation and collaboration with civil society to ensure inclusiveness and effectiveness. In particular, non-governmental organizations can act as facilitators between the government and community-based organizations.[33]

Implementation portfolio

Normative action

The Peninsula Principles also stipulate an obligation for the State to legislate, i.e. to transpose the Principles' provisions into domestic law and policies in terms of prevention, assistance and protection regarding climate displacement. One of the options in order to ensure full and effective protection of climate-displaced persons would be to transpose the Principles into a single law, rather than to include relevant provisions throughout the legal framework. For the affected communities and persons, this would also make their rights more accessible. Provisions in other laws and regulations will nevertheless need to be reviewed to ensure that climate displacement prevention, assistance and protection measures permeate the legislative framework and that procedures are in place to give effect to these provisions, including through effective remedies. Another possibility is to mainstream the provisions of the Peninsula Principles within the legal framework. An interesting example to be followed is the anchoring of climate protection in a constitutional provision. In January 2014, Tunisia became the third country, after the Dominican Republic and Ecuador, to adopt a constitution guaranteeing the right to a clean environment, thus opening the door for legislation for both the environment and climate protection at the local, regional and national level.[34]

Rights-based technical assistance

The Peninsula Principles acknowledge the burden that climate displacement settings may represent for States, in that they provide that States should in certain circumstances request technical assistance and cooperation from the international community, i.e. governments and relevant international agencies, in order to adequately prevent and respond to climate displacement.

The assistance required is certainly not limited to material or financial resources, but includes analytical and advisory services, technical expertise drawing on previous experiences and best practices,[35] as well as knowledge dissemination through institutions and commenting on legislative action. The latter is of particular importance as technical cooperation is to be undertaken under the human rights compliance proviso.

In this context, besides the involvement of international organizations, particularly UN agencies, programmes and funds as well as governmental organizations, the cooperation and support of specialized non-State actors are often instrumental, as they are playing increasingly important roles as a fundamental component of the global civil society.[36] They offer strong and direct ties with local actors, representatives of the persons concerned, community-based and grassroots organizations and, consequently, act directly at the local level and engage with the local authorities. At the same time, they are able to arrange for national, or in some instances even international, support. Besides their commitment and capacities, as well as their proximity and association with relevant actors on the ground, non-State actors exhibit a structural difference as compared to State and other actors, which may be the decisive and major asset: this is their not-for-profit and non-interest-driven foundation.[37]

Ultima ratio *relocation*

Assessing relocation needs is key to climate displacement preparedness. In order to be able to quantify future needs, assessments need to be based on expert analysis and should adopt a uniform assessment methodology and model to be applied throughout a country to ensure consistency and avoid discrepancies between regions, as well as between experts' findings and the authorities' entitlement figures.

A study of the current situation and needs of climate-displaced persons is crucial for the elaboration of a long-term strategy. Data collection and analysis need to include potential host communities as well as economic conditions and livelihoods, ethnic and demographic factors, and cultural differentiation, which are all elements determining a successful adaptation and integration process in the context of relocation. While accelerated by the change of the contextual parameters, the commitment to a rights-based approach and the centrality of human rights protection, the post-2011

Tunisia[38] example demonstrates that forecasting desertification and future relocation needs, including as a result of climate change, has become an essential element in policy formulation and implementation.[39]

The Principles' approach goes beyond post-disaster vulnerability and needs assessments. With a focus on planning, the application of the Peninsula Principles calls for pre-disaster assessments of inequalities and vulnerabilities, including to identify patterns of discrimination be they based on race, socio-economic status, tenure, gender or other relevant grounds.

Relocation is the *ultima ratio* in addressing climate displacement when return is not possible,[40] but is bound to certain conditions. As a general principle, no relocation shall take place unless individuals, households and communities provide their full and informed consent. This extends not only to those displaced but also to the host communities, recognizing that absorption capacities of communities may be limited and that a massive arrival of families may have a potentially destabilizing effect on the receiving society.

An integration programme should cover all aspects of socio-economic, cultural and educational integration as an essential part of the relocation process. The programme must take due note of local specificities in the host communities, and must devise a practice for systematic assessment and consultation in the process of selection of new host communities.

Participatory approach and information sharing

Participation and inclusiveness are other essential elements for the success of relocations. Grievances and appeals proceedings allow the identification of all interests at stake, as well as any potential difficulties and obstacles to planned relocations. The more effort is invested in this aspect of the initial process, the fewer complications are likely to arise in the course of the relocation process, in both quantity and quality. Another advantage of participatory processes is that they can utilize the innate knowledge of the local populace and communities, increase their acceptance of the process, and ensure that no false expectations associated with the relocation process are raised.

Adequate consultation can be facilitated and achieved by a constructive, participatory interactive dialogue. Such a process also builds confidence of all stakeholders, and will have a certain precluding character, as those views that were not voiced at this stage of an all-inclusive dialogue may not necessarily be taken into consideration at a later point in time.

As a general model, what can be said about consultative processes is that they ought to encompass inner and outer spheres, i.e. the internal consensus building and the external consultation. Both can consist of formal and informal components. In addition to public meetings, interested parties should be provided with an opportunity to express their views in special meetings, popular discussions and public fora. It is important that interested parties are enabled to intervene at an early stage of the process to assess the prospect for consensus building and to allow for necessary adjustments prior to implementation. Mixed committees of domestic authorities, representatives of international organizations, civil society and experts can also be very helpful in this regard. It should nevertheless be borne in mind that the emerging consensus that may be identified by means of popular consultations will often be a fragile one, which does not necessarily allow all the details of complex matters to be discussed, as this would expose the underlying and unresolved differences of opinion and conflicts. The overall value of consultative processes for decision-making, as well as for the sustainability of the outcome in terms of acceptance, abidance and implementation, cannot be emphasized enough. While the cost associated with such processes, particularly as a result of their time-intensiveness, is seemingly high, a rights-based process cannot be reduced to a simple cost-benefit analysis because it is essential for a sustainable and apprehended solution.

Proactive information collection and sharing are inherent in the participatory approach which seeks to involve the affected communities in the decision-making process. This is particularly important as often those who are most affected by disasters and the effects of climate change are also those who are not well informed about the risks and the effects of climate change and the availability of assistance, including funding.[41]

Post-relocation livelihoods

Livelihoods and basic services, including adequate housing, are critical to the success of planned relocations. It is essential that displaced and relocated persons are able to ensure their livelihoods. States should therefore, in accordance with the Peninsula Principles, adopt measures that promote livelihoods, acquisition of new skills, and economic prosperity for both displaced and host communities. Another aspect is the provision of basic services, and adequate and affordable housing and education. The Principles also define an equity standard, i.e. between relocating and host communities. Obviously, displaced persons enjoy the same rights as those who have not been relocated. It is very important that relocation does not erode the prospects of those concerned, namely by weakening social networks and diminishing cooperation due to increased competition for resources and employment, reducing livelihood and income sources and thus causing increased poverty and precarious living conditions. It is therefore necessary to understand the dynamics of vulnerability and to devise innovative social protection mechanisms to reduce the impacts of climate displacement and to ensure the protection of the most vulnerable.

Adaptation funding

Earmarking budgetary provisions to effectively resource local institutions and mechanisms dealing with climate displacement in the broadest sense (i.e. including risk management) will facilitate implementation, as lengthy discussions and negotiations about funding, which would divert from action, can be avoided. It is also necessary to stimulate investment in areas of risk reduction and to create incentives for investing in prevention. Tracking such investments will provide a clear picture of the costs and benefits involved through verifiable data and can constitute a means of further promoting implementation effectiveness.

Mechanisms for mediation and grievance redress at the local level

In implementing the Peninsula Principles, irrespective of the displacement stage, it is indispensable that mechanisms and effective remedies for mediation and grievance redress are established at the local level. Beyond these institutional mechanisms, individuals and communities also need to be put in a position to effectively claim their rights, as often those who are affected do not have the means or actual possibility to do so.

Coordination beyond silos

Coordination of domestic institutions at the national level is a key concern for the implementation of the Peninsula Principles (as well as coordination of international institutions, as outlined above). Effective application of the Principles entails overcoming the protectionist silo mentality at all levels, i.e. implementation needs to transgress institutional, thematic or organizational silos at the domestic and the international levels.

Climate displacement including risk management falls within the ambit of a number of ministries and local institutions, including the Ministry of Environmental Protection and Natural Resources, the Ministry of Agriculture, the Ministry of Finance, and the Ministry of Justice and other relevant executive bodies concerned with urban/spatial planning. These, together with regular structured civil society consultation, should oversee the formulation of a long-term strategy on climate displacement. Civil society organizations and representatives of communities displaced due to the effects of climate change and host communities should also be involved in the strategy formulation in a structured and inclusive manner. Formal consultations at several stages of the process, including at the start and before finalization, are crucial for an informed and effective policy.

In addition, the central government should work with national and international organizations and agencies, as well as with donor organizations, in order to obtain financial and informational support for the formulation and subsequent implementation of this long-term strategy.

Concluding remarks

In the light of the many existing relevant standards, first and foremost the Guiding Principles on Internal Displacement, one of the many questions raised is whether we need yet another set of Principles. The Peninsula Principles have not been crafted in a legal vacuum, but in a situation of legal gap; they explicitly build on and contextualize the Guiding Principles on Internal Displacement[42] and acknowledge, among others, the Pinheiro Principles,[43] incorporating some of their provisions.

In contrast to proposals for a new legal instrument providing protection for people displaced by climate change, the Peninsula Principles constitute a restatement of the existing law, tailoring it to the specific needs of climate displacement in order to address gaps and formulate policy and action guidance. The Principles draw on real-life scenarios and seek to formulate a common standard to provide a foundation for the articulation of the rights of those affected by climate displacement, as well as to guide the action of local and national authorities in addressing the related challenges. Moreover, in spite of the existing standards, there are protection gaps when it comes to climate displacement, both internally and across borders.

The Peninsula Principles are meant to support and guide all actors in the search for viable rights-based solutions for the climate displacement challenge that is becoming more pressing by the day. It is hoped that human rights mechanisms, national and international non-governmental organizations, as well as governments, will embrace the Principles and use them as the basis for action in support of those faced with the reality of climate change.

Meanwhile, leading by default and out of functional necessity, Displacement Solutions will continue to take on its responsibility and will act, *ad interim* and on an ad hoc basis, as registry for the Peninsula Principles. During this transitional phase, it will function as a global implementation tracking facility and continue to advocate for the appliance of the Peninsula Principles and monitor their implementation until an international actor associates itself with the process and effectively takes on the lead role.

Notes

1 http://displacementsolutions.org/category/projects/climate-displacement-law-project/, accessed 19 May 2015.
2 On the complementarity of human rights and humanitarian law, see e.g. Office of the High Commissioner for Human Rights (OHCHR), International Legal Protection of Human Rights in Armed Conflict, 2011.
3 See in this regard A/HRC/17/31, Report of the Special Representative of the Secretary-General on the issue of human rights and transnational corporations and other business enterprises, Guiding Principles on Business and Human Rights, Guiding Principles on Business and Human Rights: Implementing the United Nations 'Protect, Respect and Remedy' Framework, March 2011.

4 J.-R. Sieckmann, The Theory of Principles – A Framework for Autonomous Reasoning, in: M. Borowski (ed.), *On the Nature of Legal Principles, Proceedings of
  the Special Workshop 'The Principles Theory' held at the 23rd World Congress
  of the International Association for Philosophy of Law and Social Philosophy*,
  Krakow: Archives for Philosophy of Law and Social Philosophy, 2007,
  pp. 49–62; R. Dworkin, *Taking Rights Seriously*, Cambridge, MA: Harvard University Press, 1978, pp. 22ff.; C. Bäcker, Rules, Principles, and Defeasibility, in:
  Borowski (ed.), *On the Nature of Legal Principles*, pp. 79–103.

5 R. Zetter, Protecting People Displaced by Climate Change: Some Conceptual
  Challenges, in: J. McAdam, *Climate Change and Displacement: Multidiscipli-
  nary Perspectives*, 2012, Oxford: Hart, pp. 141ff.

6 See W. Kälin, From the Nansen Principles to the Nansen Initiative, *Forced
  Migration Review*, Issue 41, December 2012, pp. 48–49; for an assessment of
  the cross-border displacement due to the effects of climate change: W. Kälin,
  Conceptualising Climate-Induced Displacement, in: McAdam, *Climate Change
  and Displacement*, pp. 94ff.; also J. McAdam, 'Disappearing States', Statelessness and the Boundaries of International Law, in: McAdam, *Climate Change
  and Displacement*, pp. 105–129, for a legal analysis of the issue of sinking
  islands.

7 *Teitioa* v. *The Chief Executive of the Ministry of Business Innovation and
  Employment*, CIV-2013–404–3528 [2013] NZHC 3125, High Court of New
  Zealand, 26 November 2013, para. 51.

8 For a diagnosis of the rights gap, see also K. M. Wyman, Responses to Climate
  Migration, *Harvard Environmental Law Review*, Vol. 37, 2013, pp. 167–216;
  see also K. Hassine, Coping with the Realities of Climate Displacement:
  The Peninsular Principles, December 2013, https://terra0nullius.wordpress.
  com/2013/12/10/coping-with-the-realities-of-climate-displacement-the-peninsular-
  principles/#more-3908, accessed 19 May 2015.

9 E/CN.4/1998/53/Add.2, Report of the Representative of the Secretary-General,
  Guiding Principles on Internal Displacement, 1998; see also R. Cohen, Lessons
  from the development of the Guiding Principles on Internal Displacement,
  Forced Migration Review 45, February 2014, pp. 12–14.

10 A. Oliver-Smith and A. de Sherbinin, Resettlement in the Twenty-first Century,
   in: *Forced Migration Review* 45, February 2014, pp. 23–25.

11 See World Bank, Safeguards and Sustainability Policies in a Changing World:
   An Independent Evaluation of World Bank Group Experience, Independent
   Evaluation Group Study Series, 2010, http:siteresources.worldbank.org/EXT-
   SAFANDSUS/Resources/Safeguards_eval.pdf, accessed 19 May 2015; A/HRC/
   16/42/Add.2, Report of the Special Rapporteur on adequate housing as a com-
   ponent of the right to an adequate standard of living, and the right to non-
   discrimination, Preliminary note on the mission to the World Bank Group,
   2011, para. 14 et seq.; see also Oxfam/Inclusive Development International, A
   Proposal for New World Bank Safeguards on Tenure of Land, Housing and
   Natural Resources, April 2013, http://consultations.worldbank.org/Data/hab/
   files/meetings/WorldBankTenureSafeguardsSubmission_FINAL.pdf, accessed 19
   May 2015.

12 For an overview on the causality debate see, inter alia, R. Zetter, Protecting
   People Displaced by Climate Change: Some Conceptual Challenges, in:
   McAdam, *Climate Change and Displacement*, pp. 137ff.

13 See G. Hugo, Climate Change-Induced Mobility and the Existing Migration
   Regime in Asia and the Pacific, in: McAdam, *Climate Change and Displacement*, pp. 21ff.

14 E/CN.4/Sub.2/2005/17, Report of the Special Rapporteur of the Sub-
   Commission on housing and property restitution in the context of the return of

refugees and internally displaced persons, Principles on housing and property restitution for refugees and displaced persons, June 2005.

15 FAO/IDMC/OCHA/OHCHR/UN-Habitat/UNHCR, *Handbook, Housing and Property Restitution for Refugees and Displaced Persons, Implementing the 'Pinheiro Principles'*, March 2007, pp. 16ff. www.ohchr.org?Documents/Publications/pinheiro_priniciples,pdf, accessed 19 May 2015; K. Hassine, Regularizing Property Rights in Kosovo and Elsewhere, WiKu, 2009.

16 African Union Convention on the Protection of and Assistance to Internally Displaced Persons, 22 October 2009, www.unhcr.org/refworld/docid/4ae825fb2.html, accessed 19 May 2015.

17 International Institute for Sustainable Development, Land Ownership after; UN Human Settlements Programme, *Scoping Report: Addressing Land Issues After Natural Disasters*, Nairobi, UN-Habitat, 2008, pp. 6ff.; both at www.iisd.org/pdf/2006/es_addressomg_land.pdf accessed 19 May 2015.

18 See ECMI, Resettlement of Ecological Migrants in Georgia: Recent Developments and Trends in Policy, Implementation, and Perceptions, Working Paper no. 53, January 2012, pp. 3ff.; see also more generally on the situation of IDPs in Georgia, A/HRC/10/13/Add.2, Report of the Representative of the Secretary-General on the human rights of internally displaced persons, Mission to Georgia, 2009.

19 See ECMI, Resettlement of Ecological Migrants in Georgia, pp. 3ff.

20 S. Leckie, An Integrated Approach to Climate Change Demands that Human Rights and Adaptation Strategies are Pursued Hand-in-Hand, *Forced Migration Review* 31, October 2008, p. 18.

21 A/66/270, Report of the Special Rapporteur on adequate housing as a component of the right to an adequate standard of living and on the right to non-discrimination in this context to the General Assembly (on the right to adequate housing in disaster relief efforts), August 2011, para. 39 et seq.

22 Ibid., para. 53 et seq.

23 P. Kadetz, New Orleans: a Lesson in Post-disaster Resilience, in: *Forced Migration Review* 45, February 2014, pp. 61–62; A/66/270, Report of the Special Rapporteur on adequate housing as a component of the right to an adequate standard of living, paras 14, 22 and 41.

24 *Ned Comer et al.* v. *Murphy Oil USA et al.*, No. 07–60756, United States Court of Appeals for the Fifth Circuit, 22 October 2009; *Ned Comer et al.* v. *Murphy Oil USA et al.*, No. 12–60201, United States Court of Appeals for the Fifth Circuit, 14 May 2013.

25 *Budayeva and others* v. *Russia*, App. nos 15339/02, 21166/02, 20058/02, 11673/02 and 15343/02, ECHR Judgment 20 March 2008.

26 *Budayeva and others* v. *Russia*, App. nos 15339/02, 21166/02, 20058/02, 11673/02 and 15343/02, ECHR Judgment 20 March 2008.

27 UN Human Settlements Programme, Scoping Report, pp. 30ff.; S. Badri, Ali Asgary, A.R. Eftekhari and Jason Levy, Post-disaster Resettlement, Development and Change: a Case Study of the 1990 Manjil Earthquake in Iran, *Disasters* 30 (4), pp. 451–468.

28 In this regard see also various disaster law initiatives, particularly the International Disaster Response Law Guideline developed by the International Federation of Red Cross and Red Crescent Societies, https://www.ifrc.org/what-we-do/disaster-law/about-disaster-law/international-disaster-response-laws-rules-and-principles/idrl-guidelines/, accessed 19 May 2015.

29 See C. Bäcker, Rules, Principles, and Defeasibility, pp. 79–103.

30 J. Donnelly, *International Human Rights*, Philadelphia, PA: Westview Press, 2013, pp. 81ff.

31 See e.g. Secretary-General's remarks on Climate Change at the Policy Spotlight with Friends of Europe, 3 April 2014, www.un.org/sg/statements/index.asp?nid=7563 accessed 19 May 2015.

32 A/HRC/17/31, Report of the Special Representative of the Secretary-General on the issue of human rights and transnational corporations and other business enterprises.

33 Habitat International Coalition, *Building the City with the People, New Trends in Community Initiatives in Cooperation with Local Government*, Mexico City: Habitat International Coalition, 1997, pp. 47ff.

34 Article 45: L'Etat garantit le droit à un environnement sain et équilibré et la participation à la sécurité du climat. L'Etat se doit de fournir les moyens nécessaires à l'élimination de la pollution environnementale, www.marsad.tn/uploads/documents/Constitution_Tunisienne_en_date_du_26–01–2014_Version_Francaise_traduction_non_officielle_Al_Bawsala.pdf, accessed 19 May 2015; see also a new study on the interplay between climate change, food prices and politics that fuel the Arab Spring, Center for American Progress, The Arab Spring and Climate Change, A Climate and Security Correlations Series, February 2013, www.americanprogress.org/issues/security/report2013/02/28/54579/the-arab-spring-and-climate-change, accessed 19 May 2015.

35 In particular in relation to resettlement/relocation, both so-called development-forced displacement and resettlement and post-disaster resettlement/relocation, there are a number of important lessons learned. See inter alia PPLA/2012/04, E. Ferris, Protection and Planned Relocation in the Context of Climate Change, UNHCR Legal and Protection Policy Research Series, Division of International Protection, August 2012, pp. 14ff.; Hugo, Climate Change-Induced Mobility and the Existing Migration Regime in Asia and the Pacific, pp. 26ff.; R. Bronen, Choice and Necessity: Relocations in the Arctic and South Pacific, *Forced Migration Review* 45, February 2014, pp. 17–21; J. Campbell, Climate-Induced Community Relocation in the Pacific: The Meaning and Importance of Land, in: McAdam, *Climate Change and Displacement*, pp. 57–79.

36 M. Edwards, *Civil Society*, Cambridge, UK, and Malden, MA: Policy Press, 2009, p. 63 et seq.

37 Habitat International Coalition, Building the City with the People, pp. 47ff.

38 A. Honwana, *Youth and Revolution in Tunisia*, London and New York: Zed Books, 2013, pp. 71ff.

39 See for instance the work of the Institution des Régions Arides, www.ira.agrinet.tn/ang/, and of the Association pour le Développement Durable.

40 In determining the feasibility of return, the question of durable solution needs also to be examined: see UNEP, Sudan, Post-Conflict Environmental Assessment, 2007, pp. 115–117, www.unep.org/sudan/post-conflict/PDF/UNEP-sudan.pdf, accessed 19 May 2015.

41 See also Convention on Access to Information, Public Participation in Decision-Making and Access to Justice in Environmental Matters, 25 June 1998.

42 E/CN.4/1998/53/Add.2, Report of the Representative of the Secretary-General, Guiding Principles on Internal Displacement, 1998.

43 E/CN.4/Sub.2/2005/17, Report of the Special Rapporteur of the Sub-Commission on housing and property restitution.

10 Some observations and conclusions

Scott Leckie

Since the adoption of the Peninsula Principles on 18 August 2013, they have received extensive and growing attention from climate change experts, lawyers, UN agencies, governments, grassroots groups, NGOs and communities that are already enduring climate displacement. The Principles have now been translated into eight languages (with at least six more translations to come) and applied to a growing number of climate displacement cases in Australia, Bangladesh, Fiji, Panama, the Solomon Islands, the US and Vanuatu. They have been discussed in innumerable meetings across the world and included in a growing number of official reports, books and other publications. The Principles have been taught in some of the world's leading law schools and shared with hundreds of government officials from more than 50 countries. They have been the subject of reports on climate displacement, and have been cited in an array of publications and websites. Clearly, the Principles have arrived, and have shown themselves to be of growing utility for those engaged in practical rights-based efforts to manage climate displacement.

At the same time, of course, we have proverbial light years to go before we reach the stage where States, UN organs, civil society actors and communities themselves in practice treat – in the terms of the Peninsula Principles – climate-displaced persons (CDPs), households and communities as rights-holders: specific subjects of human rights law and the clear beneficiaries of specialised laws, policies, institutions and programmes designed to protect their assets and their housing, land and property (HLP) and other rights. Political entities that wish to address the rights of CDPs, in preventive, protective and remedial senses, now have the normative framework provided for within the Peninsular Principles as a clear and consolidated basis for starting processes *now* to address this increasingly serious challenge.

Experts from NGOs such as Displacement Solutions have worked in and examined in detail the climate displacement that is already well under way in many countries. From the testimony of the communities affected, combined with frequent and ongoing discussions with responsible government officials in these nations, it is abundantly clear that the vast majority of those who are being and will be displaced by the forces of climate change will choose to remain in their own countries. Certainly, some will seek the protection of third countries and attempt to find new migration options in countries other than their own, but it is beyond dispute that the majority of the scores or hundreds of millions of people who will move will remain within their countries of origin. This fact was instrumental in guiding the manner by which the Peninsular Principles were developed and is grounded deeply in repeated observations on the ground in all regions of the world. These facts are not based on one-off, so-called 'consultations' held in capital cities (often far from where climate displacement is occurring) that last for two days in luxury hotels largely attended by state officials, but are rather built on years of field work with local organisations and communities (work that will continue for years to come, as well!), coupled with progressive readings of international human rights and other norms that begin and end with *Principles*, not conservative 'pragmatism' dictated by the reluctance of States to do more for the world's climate-displaced populations. Many of those who take a global view of climate displacement and monitor how international institutions are addressing this issue are rather startled to see so much attention paid to 'cross-border' displacement and migration, when such forms of climate displacement are clearly a very small proportion of the total number of people who are and will be affected by environmental processes. While those working on these issues are certainly happy to see that governmental and other donors are beginning to devote interest and resources to the question at hand, there is nonetheless some dissatisfaction that so much of this attention has focused on such a small fraction of the true nature and scale of the problem.

Why, when perhaps 95 per cent or more of those being displaced and who will be displaced will likely remain within the borders of their countries, have many donors chosen to focus attention on the international movements and migration of the comparatively limited proportion of CDPs who will seek solutions far from home? We can in part understand this approach because of the clear legal gap that exists for CDPs who cross borders and the inadequacies of existing global standards to address these concerns. But this is not the whole story.

Far more central to understanding why the financial and, therefore, political support thus far has tended to focus on cross-border movements and international migration, I believe, is the ever-growing reticence of governments the world over to apply human rights norms – which they have already in most instances freely agreed to comply with – to socially, economically and politically weaker segments of society. Taking rights seriously and protecting the most vulnerable is hard for many States to accept, and the mechanisms to enforce these rights and secure redress when they are infringed are all too frequently lacking. While there are, of course, instances where these types of rights protections, including HLP rights protections, have expanded and been accorded growing seriousness in recent years, the larger trend has been towards ideologically driven, essentially neo-liberal social and economic policies that rely on the market and individual effort without State support to solve whatever social or economic challenge a government may face. This, in turn, has driven donors concerned about climate displacement right into the arms of those agencies that in their very essence take precisely the same approach: 'We know governments are reluctant to assist, so let's allow individual responsibility and migration to greener pastures guide us on our way in addressing what is clearly one of the most severe *human rights* challenges the world will face in the twenty-first century.' Is this really the best the world has to offer?

The Peninsula Principles provide a way out of this minimalist approach and offer a normative pathway that is grounded in the inherent rights of everyone displaced or to be displaced by climate change. They are built upon the basic foundations of existing international human rights law and provide an evidence- and practice-based conceptualisation of how CDPs can best be assured that all of their human rights will be fully respected, protected and fulfilled. At their core, the Principles assert that neither migration nor the market alone will be sufficient if we are to take the rights of CDPs seriously.

We need a much more all-encompassing approach, one that engages all of the actors required, a truly comprehensive approach to the rights of CDPs; an approach that strives first to prevent climate displacement wherever possible by ensuring that measures of adaptation are designed specifically to enable people to stay where they are for as long as they wish and as long as it is viable to do so, and when climate displacement is inevitable (now or in the future) an approach that engages fundamentally with the people affected and ensures that human rights protections are at the core of all responses to climate displacement. The Peninsula Principles, of course, are neither a panacea nor the only standards upon which good and new law and policy can be based, but they are important standards, that all of us who support them hope will be increasingly used as the basis for finding positive ways forward on these very difficult issues.

The world stands on the brink of what will likely be the largest mass displacement of people and communities ever registered in recorded history. And yet, governments, the UN, and even the human rights activist community as of 2015 all remain dreadfully unprepared for the large-scale movements of people to come. Much of the clamour made thus far about this issue has sought to rely on individual migration solutions to the growing climate displacement crisis, in the mistaken belief that simply prising open various migration pathways is an adequate approach to an issue which at its core is one of human rights. Such *laissez-faire* approaches, though, fall short of what is required to protect not just the individuals and households, but the *communities* – villages, settlements, towns – that are sadly positioned in such a way that their fate has become increasingly known. As a process and as a logical choice, migration for individuals and households to ostensibly better horizons has already – long before climate change came around – had a major influence on the shape of societies the world over and has for many, resulted in better and freer lives. As such, proponents of a migration approach are in effect leaving solutions up to individual choice and entrepreneurship, which may suit some but will never suit everyone.

A more profound problem, however, is the fatal flaw of such a minimalist approach to climate displacement: this is simply that it ignores the fundamentally unique quality of climate displacement when compared to other forms of forced displacement and coerced movement, in that it does and will primarily affect groups of people, or communities, with wishes, dynamics and histories far larger and more complex than whatever individual migration choices someone may choose to make. Far more often than not, when confronted with looming crises, communities forced to move wish to stay together and move as a group to a new and safer location.

Migration minimalists may claim that this is merely another manifestation of migration, but in fact it is far from that. When given the option, especially while engaging with active and responsible States, communities that have no other option than flight inevitably prefer to move in a planned, coordinated and self-driven manner and be relocated according to an agreed location, equal to or better than their current place of abode. Indeed, such community-level planned relocation processes are already under way in response to climate impacts in Bangladesh, Panama, Papua New Guinea, the Solomon Islands and elsewhere. This is what the people within communities want, and documents like the Peninsula Principles provide the normative framework which can make this approach as successful as possible.

Migration approaches are based on a position that communities will not wish to move as a group, that governments or international agencies will be incapable of coordinating adequate planned relocation efforts, and ultimately that human rights are peripheral issues to these processes, rather than the central core of a multi-faceted approach that actively seeks to protect CDPs and provide them with the best possible outcome to the tragedy that is climate displacement. One way in which these rights-based approaches can best be pursued is reliance by States, the UN, NGOs and the people themselves the world over on the guidance provided by the Peninsula Principles.

Appendix

The Peninsula Principles on Climate Displacement within States (18 August 2013)

Table of contents

Preamble

CONCERNED that events and processes caused or exacerbated by climate change have and will continue to contribute to displacement of populations resulting in the erosion of the rights of those affected, in particular vulnerable and marginalized groups, the loss of assets, housing, land, property and livelihoods, and the further loss of cultural, customary and/or spiritual identity;

GUIDED by the Charter of the United Nations, and REAFFIRMING the Universal Declaration of Human Rights, the International Covenant on Economic, Social and Cultural Rights, the International Covenant on Civil and Political Rights as well as the Vienna Declaration and Program of Action;

NOTING that these Peninsula Principles build on and contextualize the United Nations Guiding Principles on Internal Displacement to climate displacement within States;

UNDERSTANDING that when an activity raises threats of harm to human health, life or the environment, precautionary measures should be taken;

COGNISANT that the vast majority of climate displaced persons are not responsible for the processes driving climate change;

NOTING that while climate displacement can involve both internal and cross-border displacement, most climate displacement will likely occur within State borders;

REAFFIRMING the right of climate displaced persons to remain in their homes and retain connections to the land on which they live for as long as possible, and the need for States to prioritise appropriate mitigation, adaptation and other preventative measures to give effect to that right;

REAFFIRMING further the right of those who may be displaced to move safely and to relocate within their national borders over time;

RECOGNISING that voluntary and involuntary relocation often result in the violation of human rights, impoverishment, social fragmentation and other negative consequences, and recognising the imperative to avoid such outcomes;

NOTING further that climate displacement if not properly planned for and managed may give rise to tensions and instability within States;

ACKNOWLEDGING that States bear the primary responsibility for their citizens and others living within their jurisdiction, but recognising that, for many States, addressing the issue of and responding to climate displacement presents financial, logistical, political, resource and other difficulties;

CONVINCED, however, that as climate change is a global problem, States should, on request by affected States, provide adequate and appropriate support for mitigation, adaptation, relocation and protection measures, and to provide assistance to climate displaced persons;

REALISING that the international community has humanitarian, social, cultural, financial and security interests in addressing the problem of climate displacement in a timely, coordinated and targeted manner;

REALISING further that there has been no significant coordinated response by States to address climate displacement, whether temporary or permanent in nature;

RECOGNISING that the United Nations Framework Convention on Climate Change (UNFCCC) and its Kyoto Protocol neither contemplate nor address the issue of climate displacement, and that conferences and meetings of the parties to these instruments have not substantively addressed climate displacement other than in the most general of terms;

NOTING, however, that paragraph 14(f) of the UNFCCC 16th session of the Convention of the Parties (COP16) Cancún Adaptation Framework refers to enhanced action on adaptation, including '[m]easures to enhance understanding, coordination and cooperation with regard to climate change induced displacement, migration and planned relocation...';

NOTING further that UNFCCC COP18 in Doha decided to establish, at UNFCCC COP19, institutional arrangements to address loss and damage associated with climate change impacts in developing countries that are particularly vulnerable to the adverse effects of climate change as part of the Cancún Adaptation Framework;

RECOGNISING the work being undertaken by the United Nations and other inter-governmental and non-governmental agencies to address climate displacement and related factors;

REALISING the need for a globally applicable normative framework to provide a coherent and principled approach for the collaborative provision of pre-emptive assistance to those who may be displaced by the effects of climate change, as well as effective remedial assistance to those who have been so displaced, and legal protections for both;

ACKNOWLEDGING the IASC Operational Guidelines on the Protection of Persons in Situations of Natural Disasters, the Hyogo Framework for Action, the UN Principles on Housing and Property Restitution for Refugees and Displaced Persons and others, the incorporation of a number of their Principles within these Peninsula Principles, and their application to climate displaced persons;

ACKNOWLEDGING also regional initiatives addressing internal displacement such as the African Union Convention for the Protection and Assistance of Internally Displaced Persons in Africa;

NOTING the work of the Nansen Initiative on disaster-induced cross-border displacement;

NOTING that these Peninsula Principles, addressing climate displacement within States, necessarily complement other efforts to address cross-border displacement; and

RECOGNISING judicial decisions and the writings of eminent jurists and experts as a source of international law, and acknowledging their importance and contribution to formulating the present Peninsula Principles, these Peninsula Principles on Climate Displacement ('Peninsula Principles') provide as follows:

I Introduction

Principle 1: Scope and purpose

These Peninsula Principles:

a   provide a comprehensive normative framework, based on Principles of international law, human rights obligations and good practice, within which the rights of climate displaced persons can be addressed;
b   address climate displacement within a State and not cross-border climate displacement; and
c   set out protection and assistance Principles, consistent with the UN Guiding Principles on Internal Displacement, to be applied to climate displaced persons.

Principle 2: Definitions

For the purposes of these Peninsula Principles:

a   'Climate change' means the alteration in the composition of the global atmosphere that is in addition to natural variability over comparable time periods (as defined by the IPCC).
b   'Climate displacement' means the movement of people within a State due to the effects of climate change, including sudden and slow-onset environmental events and processes, occurring either alone or in combination with other factors.
c   'Climate displaced persons' means individuals, households or communities who are facing or experiencing climate displacement.
d   'Relocation' means the voluntary, planned and coordinated movement of climate displaced persons within States to suitable locations, away from risk-prone areas, where they can enjoy the full spectrum of rights including housing, land and property rights and all other livelihood and related rights.

Principle 3: Non-discrimination, rights and freedoms

a   States shall not discriminate against climate displaced persons on the basis of their potential or actual displacement, and should take steps to repeal unjust or arbitrary laws and laws that otherwise discriminate against, or have a discriminatory effect on, climate displaced persons.
b   Climate displaced persons shall enjoy, in full equality, the same rights and freedoms under international and domestic law as do other persons in their country, in particular housing, land and property rights.
c   States should ensure that climate displaced persons are entitled to and supported in claiming and exercising their rights and are provided with effective remedies as well as unimpeded access to the justice system.

Principle 4: Interpretation

a   These Peninsula Principles shall not be interpreted as limiting, altering or otherwise prejudicing rights recognised in international law, including human rights, humanitarian law and related standards, or rights consistent with those laws and standards as recognised under domestic law.

b   States should interpret these Peninsula Principles broadly, be guided by their humanitarian purpose, and display fairness, reasonableness, generosity and flexibility in their interpretation.

II General obligations

Principle 5: Prevention and avoidance

States should, in all circumstances, comply in full with their obligations under international law so as to prevent and avoid conditions that might lead to climate displacement.

Principle 6: Provision of adaptation assistance, protection and other measures

a   States should provide adaptation assistance, protection and other measures to ensure that individuals, households and communities can remain in their homes or places of habitual residence for as long as possible in a manner fully consistent with their rights.

b   States should, in particular, ensure protection against climate displacement and demonstrate sensitivity to those individuals, households and communities within their territory who are particularly dependent on and/or attached to their land, including indigenous people and those reliant on customary rules relating to the use and allocation of land.

Principle 7: National implementation measures

a   States should incorporate climate displacement prevention, assistance and protection provisions as set out in these Peninsula Principles into domestic law and policies, prioritising the prevention of displacement.

b   States should immediately establish and provide adequate resources for equitable, timely, independent and transparent procedures, institutions and mechanisms – at all levels of government (local, state and national) – to implement these Peninsula Principles and give effect to their provisions through specially earmarked budgetary allocations and other resources to facilitate that implementation.

c   States should ensure that durable solutions to climate displacement are adequately addressed by legislation and other administrative measures.

d   States should ensure the right of all individuals, households and communities to adequate, timely and effective participation in all stages of policy development and implementation of these Peninsula Principles, ensuring in particular such participation by indigenous peoples, women, the elderly, minorities, persons with disabilities, children, those living in poverty, and marginalized groups and people.

e   All relevant legislation must be fully consistent with human rights laws and must in particular explicitly protect the rights of indigenous peoples, women, the elderly, minorities, persons with disabilities, children, those living in poverty, and marginalized groups and people.

Principle 8: International cooperation and assistance

a   Climate displacement is a matter of global responsibility, and States should cooperate in the provision of adaptation assistance (to the maximum of their available resources) and protection for climate displaced persons.

b   In fulfilling their obligations to prevent and respond to climate displacement within their territory, States have the right to seek cooperation and assistance from other States and relevant international agencies.

c   States and relevant international agencies, either separately or together, should provide such cooperation and assistance to requesting States, in particular where the requesting State is unable to adequately prevent and respond to climate displacement.

d   States that are otherwise unable to adequately prevent and respond to climate displacement should accept appropriate assistance and support from other States and relevant international agencies, whether made individually or collectively.

III Climate displacement preparation and planning

Principle 9: Climate displacement risk management

States, in terms of climate displacement risk management, monitoring and modeling, using a rights-based approach, should:

a   identify, design and implement risk management strategies, including risk reduction, risk transfer and risk sharing mechanisms, in relation to climate displacement;

b   undertake systematic observation and monitoring of, and disaggregated data collection at the household, local, regional and national levels on, current and anticipated climate displacement;

c   enhance sharing, access to and the use of such data at the household, local, regional and national levels, mindful of the need for data protection and predetermined use of data, facilitate the assessment and management of climate displacement;

d model likely climate displacement scenarios (including timeframes and financial implications), locations threatened by climate change, and possible relocation sites for climate displaced persons;

e integrate relocation rights, procedures and mechanisms, as defined in these Peninsula Principles, within national laws and policies; and

f develop institutional frameworks, procedures and mechanisms with the participation of individuals, households and communities that:

> i identify indicators that will, with as much precision as possible, classify where, at what point in time, and relevant to whom, relocation will be required as a means of providing durable solutions to those affected;
>
> ii require and facilitate governmental technical assistance and funding; and
>
> iii outline steps individuals, households and communities can take prior to climate displacement in order to receive such technical assistance and financial support.

Principle 10: Participation and consent

To enable successful preparation and planning for climate displacement, States should:

a ensure that priority consideration is given to requests from individuals, households and communities for relocation;

b ensure that no relocation shall take place unless individuals, households and communities (both displaced and host) provide full and informed consent for such relocation;

c only require relocation to take place without such consent in exceptional circumstances when necessary to protect public health and safety or when individuals, households and communities face imminent loss of life or limb;

d adopt measures that promote livelihoods, acquisition of new skills, and economic prosperity for both displaced and host communities;

e make certain that:

> i affected individuals, households and communities (both displaced and host) are fully informed and can actively participate in relevant decisions and the implementation of those decisions, including the planning and implementation of laws, policies and programs designed to ensure respect for and protection of housing, land and property rights;
>
> ii basic services, adequate and affordable housing, education and access to livelihoods (without discrimination) will be available for

climate displaced persons in the host community at a standard ensuring equity between the host and relocating communities, and consistent with the basic human rights of each;

iii adequate mechanisms, safeguards and remedies are in place to prevent and resolve conflicts over land and resources; and

iv the rights of individuals, households and communities are protected at all stages of the relocation process;

f   prior to any relocation, prepare a master relocation plan that addresses critical matters including:

   i land acquisition;
   ii community preferences;
   iii transitional shelter and permanent housing;
   iv the preservation of existing social and cultural institutions and places of climate displaced persons;
   v access to public services;
   vi support needed during the transitional period;
   vii family and community cohesion;
   viii concerns of the host community;
   ix monitoring mechanisms; and
   x grievance procedures and effective remedies.

Principle 11: Land identification, habitability and use

a   Recognising the importance of land in the resolution of climate displacement, States should:

   i identify, acquire and reserve sufficient, suitable, habitable and appropriate public and other land to provide viable and affordable land-based solutions to climate displacement, including through a National Climate Land Bank;
   ii develop fair and just land acquisition and compensation processes and appropriate land allocation programmes, with priority given to those most in need; and
   iii plan for and develop relocation sites including new human settlements on land not at risk from the effects of climate change or other natural or human hazards and, in so planning, consider the safety and environmental integrity of the new site(s), and ensure that the rights of both those relocated and the communities that host them are upheld.

b   In order to determine the habitability and feasibility of any relocation site, and to ensure that climate displaced persons being relocated and the relevant jurisdictional authority are in agreement as to the habitability of any such site, States should create and make publicly available specific, geographically appropriate, standard criteria including:

    i current and future land use;

    ii restrictions (including those of a customary nature or not otherwise formally codified) associated with the land and its use;

    iii habitability of the land, including issues such as accessibility, availability of water, vulnerability to climate or other natural or human hazards, and use; and

    iv feasibility of subsistence/agricultural use, together with mechanisms for climate displaced persons to decide to where they wish to voluntarily relocate.

c   States should provide easily accessible information to individuals, households and communities concerning:

    i the nature and extent of the actual and potential changes to the habitability of their homes, lands and places of habitual residence on which they dwell or subsist, resulting from climate change, including the evidence on which such assessments are made;

    ii evidence that all viable alternatives to relocation have been considered, including mitigation and adaptation measures that could be taken to enable people to remain in their homes and places of habitual residence;

    iii planned efforts to assist climate displaced persons in relocation;

    iv available compensation and alternative relocation options if the relocation site offered is unacceptable to climate displaced persons; and

    v rights under international and domestic law, in particular housing, land and property rights.

d   States should include in relocation planning:

    i measures to compensate climate displaced persons for lost housing, land and property;

    ii assurances that housing, land, property and livelihood rights will be met for all climate displaced persons, including those who have informal land rights, customary land rights, occupancy rights or rights of customary usage, and assurances that such rights are ongoing; and

    iii assurances that rights to access traditional lands and waters (for example, for hunting, grazing, fishing and religious purposes) are maintained or similarly replicated.

Principle 12: Loss and damage

States should develop appropriate laws and policies for loss suffered and damage incurred in the context of climate displacement.

Principle 13: Institutional frameworks to support and facilitate
the provision of assistance and protection

a States should strengthen national capacities and capabilities to identify and address the protection and assistance needs of climate displaced persons through the establishment of effective institutional frameworks and the inclusion of climate displacement in National Adaptation Programmes of Action as appropriate.

b States should take all appropriate administrative, legislative and judicial measures, including the creation of adequately funded Ministries, departments, offices and/or agencies at the local (in particular), regional and national levels empowered to develop, establish and implement an institutional framework to:

  i enable government technical assistance and funding to prevent, prepare for and respond to climate displacement;
  ii support and facilitate the provision of assistance and protection to climate displaced persons;
  iii exchange information and cooperate with indigenous peoples, women, the elderly, minorities, persons with disabilities, children, those living in poverty, and marginalized groups;
  iv represent the needs of climate displaced persons.

c Responsibility for establishing Ministries, departments, offices and/or agencies should lie with national governments, and such governments should consult and collaborate with regional and local authorities, and integrate such Ministries, departments, offices and/or agencies in relevant institutional frameworks.

d States should ensure the provision of adequate resources (including points of contact and assistance) at all levels of government that directly address the concerns of climate displaced persons.

IV Displacement

Principle 14: State assistance to those climate displaced persons
experiencing displacement but who have not been relocated

a States have the primary obligation to provide all necessary legal, economic, social and other forms of protection and assistance to those climate displaced persons experiencing displacement but who have not been relocated.

b Protection and assistance activities undertaken by States should be carried out in a manner that respects both the cultural sensitivities prevailing in the affected area and the Principles of maintaining family and community cohesion.

c   States should provide climate displaced persons experiencing displacement but who have not been relocated with a practicable level of age and gender-sensitive humanitarian assistance including, without limitation, as the context requires:

   i   emergency humanitarian services;
   ii   evacuation and temporary and effective permanent relocation;
   iii   medical assistance and other health services;
   iv   shelter;
   v   food;
   vi   potable water;
   vii   sanitation;
   viii   measures necessary for social and economic inclusion including, without limitation, anti-poverty measures, free and compulsory education, training and skills development, and work and livelihood options, and issuance and replacement of lost personal documentation; and
   ix   facilitation of family reunion.

Principle 15: Housing and livelihood

a   States should respect, protect and fulfil the right to adequate housing of climate displaced persons experiencing displacement but who have not been relocated, which includes accessibility, affordability, habitability, security of tenure, cultural adequacy, suitability of location, and non-discriminatory access to basic services (for example, health and education).

b   Where climate displacement results in the inability of climate displaced persons to return to previous sources of livelihood, appropriate measures should be taken to ensure such livelihoods can be continued in a sustainable manner and will not result in further displacement, and opportunities created by such measures should be available without discrimination of any kind.

Principle 16: Remedies and compensation

Climate displaced persons experiencing displacement but who have not been relocated and whose rights have been violated shall have fair and equitable access to appropriate remedies and compensation.

V Post-displacement and return

Principle 17: Framework for return

a   States should develop a framework for the process of return in the event that displacement is temporary and return to homes, lands or places of habitual residence is possible and agreed to by those affected.

b   States should allow climate displaced persons experiencing displacement to voluntarily return to their former homes, lands or places of habitual residence, and should facilitate their effective return in safety and with dignity, in circumstances where such homes, lands or places of habitual residence are habitable and where return does not pose significant risk to life or livelihood.

c   States should enable climate displaced persons to decide on whether to return to their homes, lands or places of habitual residence, and provide such persons with complete, objective, up-to-date and accurate information (including on physical, material and legal safety issues) necessary to exercise their right to freedom of movement and to choose their residence.

d   States should provide transitional assistance to individuals, households and communities during the process of return until livelihoods and access to services are restored.

VI Implementation

Principle 18: Implementation and dissemination

States, who have the primary obligation to ensure the full enjoyment of the rights of all climate displaced persons within their territory, should implement and disseminate these Peninsula Principles without delay and cooperate closely with inter-governmental organisations, non-government organisations, practitioners, civil society, and community-based groups toward this end.

Adopted by a group of eminent jurists, text writers, legal scholars and climate change experts in Red Hill on the Mornington Peninsula, Victoria, Australia on 18 August 2013.

Index